IRISH RIVERS

edited by
Éamon de Buitléar

with contributions by
Professor James Dooge
Roger Goodwillie
Dr Hester Heuff
Dr Ken Whelan

This book is supported by Cement Roadstone Holdings plc

ACKNOWLEDGEMENTS

The publishers would like to acknowledge the advise and the assistance of the following people in the preparation of this book.

Professor Frank Mitchell author of the *Irish Landscape*.
Dr Paul Toner of An Foras Forbartha, Dublin.
Dr W. Watts, Trinity College Dublin.

First published in 1985 by Country House.
Country House is an imprint of Amach Faoin Aer Publishing
2 Cambridge Villas, Rathmines, Dublin 6, Ireland.

General Editors	Éamon de Buitléar and Roger Goodwillie
Managing Editor	Treasa Coady
Text Editor	Siobhán Parkinson
Photography	Richard T. Mills, Liam Blake, Éamon de Buitléar
Drawings by	Bob Quinn
Designed by	Wendy Dunbar
Photographic selection	Treasa Coady
Typesetting by	Glynis Millar

Cataloguing in Publication Data
1. Rivers – Ireland
I. de Buitléar, Éamon II Series
551.4'83'09415 GB 12907
ISBN 0 946172 05 6

Printed by Criterion Press, Dublin 11
Distributed by Easons of Dublin

Contents

CONTRIBUTORS

Éamon de Buitléar is an independent wildlife film maker and has been making films for RTE, Ireland's national television station, for twenty-five years. More recently his work for the BBC, noteably his most challenging film *Wild Ireland* has been widely praised. He has written many children's books on wildlife and has edited a best selling book, also called *Wild Ireland*

Jim Dooge was brought up in Dún Laoghaire in County Dublin and educated at Christian Brothers Schools Dún Laoghaire, University College Dublin, and the University of Iowa. He is known internationally for his contributions to the development of new thinking in hydrology and for his promotion of international co-operation in science and engineering. He is past president of the Institution of Engineers of Ireland and of the International Association for Hydrological Sciences, former secretary of the Royal Irish Academy and former secretary general of the International Council for Scientific Unions, foreign member of the Polish Academy of Sciences and of the Accademia Patavina (Padua) and winner of the International Prize for Hydrology for 1983.

Roger Goodwillie is a botanist and bird watcher who has been interested in rivers since childhood. He took his primary degree in Trinity College Dublin where he developed a keen interest in peatlands. He received his masters degree in the University of Toronto. He has canoed on several rivers in Canada and Ireland and now works as an ecologist with An Foras Forbartha.

Hester Heuff was born in 1947 in the Netherlands and came to live in Ireland in 1971. She received her doctoral degree from Amsterdam University in 1974 for work on the ecology of blanket bogs. She has worked on an ecological study of the aquatic plants of the Caragh river County Kerry and she now carries out botanical surveys of lakes, rivers and wetlands for the Forest and Wildlife Service of the Department of Fisheries and Forestry in Ireland. She has a keen interest in environmental planning, nature conservation and aquatic ecology.

Ken Whelan has been a keen angler since the age of five. He is one of those lucky people whose hobby has also become their profession. He joined the Inland Fisheries Trust as a research scientist in 1975 and worked on the biology of Irish mayflies, the subject of his PhD thesis. He has carried out research into the effects of peat bog development on the river Suck catchment and is now advisor to the Central Fisheries Board on all matters relating to the management, exploitation and development of salmon and sea trout stocks.

Introduction

The County brook flowed past our bedroom window and the river Dargle, which was on the other side of the house, ran only a few yards beyond the kitchen. Growing up in the Dargle valley meant that I always had the sound of a waterfall in my ears, a lullaby without which I could hardly sleep. Most of the music came from the weir, a most magical place for wildlife.

This was where I grew up with my brothers and sisters and it was on the banks of the river that we spent most of our playing time. My mother used to say that she should have grown a neck like a swan, from all the craning she did, watching to see if any of her brood had fallen into the water!

Above the weir, the river was flat and slow-moving. It seemed to hesitate before crossing this solid concrete barrier and then rushed down the moss-covered slope into two large fish pools below. Waterhens, wild duck and kingfishers nested along the flats above the tumbling cascade. A patient heron stood forever motionless at the edge. He was part of the landscape. He waited and waited at this spot, where some unsuspecting eel or trout with an urge to find new ground further upstream would make the supreme effort of trying to swim up the fall. Quick as lightning, the grey statue would come to life as the victim was grabbed from the water by a rapier-like beak and swallowed head first.

This part of the Dargle, where we lived, was a meeting place for a great variety of birds, insects, fish and mammals. The tumbling water suited many small aquatic animals, it had food for both fish and birds, and for the mammals it provided a crossing place. There were live wildlife programmes to be seen through our window everyday. These sightings of animals were my first contact with what were to be many future friends.

In his spare time, my father fly fished for trout and it was only natural that I, as a five-and-a-half-year-old, would want to do the same. I was duly supplied with a home-made rod on a particular day when the river was in flood, conditions which were more suitable for bait fishing than for the art of casting the artificial fly. A lowly worm was unceremoniously impaled for me on a small hook and I took up my position at the edge of the weir pool. This meant displacing the heron, who left hurriedly, squawking in protest as he flapped his way upstream. One could have felt sorry for him as he seemed to hold a season ticket for this stand on the Dargle.

I imitated the heron by standing as he did, staring into the water, a small figure in bare feet with a long rod clutched in tiny hands trying to stand still. It must have been beginner's luck, because a sizeable brown trout a pound in weight grabbed my bait. Knowing nothing of what was happening down below me in the darkness of the murky pool, I attempted to lift my rod for another cast. It felt heavy and as I pulled and pulled again, suddenly the large trout came splashing and kicking and hurtling towards me. It was too much. With a gasp and a scream I fled! In my wild panic I forgot to drop my rod, so instead of leaving the monster behind it followed me as I tried to run backwards up the dry slope of the weir. Whatever disaster lay in store for this budding Isaac Walton and his catch was luckily averted by my father arriving just in time to save both boy and fish!

Anglers make good naturalists, as they spend many hours wandering along river banks or silently floating their boats on our many waterways. In those situations they can observe much of the plant and animal life around them. Without the watchful eyes of these fishermen, many more Irish lakes and rivers would be lost to pollution.

The weir pool attracted quite a number of angling characters, including some local poachers. These County Wicklow fishermen were my first introduction to the many hidden secrets of the river.

There was Johnny the motor mechanic and little Ned from Kilkenny. Johnny had a smiling face and curly dark hair. He always seemed black and oily, as if he had just left off repairing some engine before coming to fish. His long waders had lovely patches of the kind I had seen vulcanised on motor car tubes. Ned was thin, and as wiry as a whippet. He seemed to have an endless supply of check tweed caps and I never saw him without this essential piece of headgear, which was really so much a part of him. Ned used to wonder at Johnny's talent as a tier of the most beautiful and delicate of artificial flies. These were painstakingly made from carefully chosen feathers and tinsels, silken threads and tiny pieces of animal furs. 'How can a man with spades for hands, fashion such tiny works of arts?' would be Ned's comment to another angler, as Johnny, who had shown his flies to Ned, made his way upstream to tempt some big trout with his latest insect imitations.

Ned was a skilled tier himself and he specialised in making colourful low-water salmon flies which were a joy for any fisherman to have in his tackle box. He had the biggest library of angling literature in the county and the biggest collection of witty stories on the river. Other piscatorial characters on the Dargle included the poacher Harvey and Hubert the gipsy. Hubert lived in a horse-drawn caravan, close to the

Hard rock faces can often result in steep and spectacular waterfalls. Without man's interventio[n] these often prove an impossible obstacle to all migrating fish. These falls [a]t Ennistimon on the Cullenag[h] or Inagh river is a typical example of such a physical barrier.
photograph: Liam Blake

Overleaf
Water flowing over the we[ir] near Cahir in County Tipperary entrains air thus producing the characteristi[c] white water and enhancing the capacity of the river to assimilate minor pollution.
photograph: Liam Blake

LOCKE'S
WHISKEY

river. He was olive skinned, and his accent was strange to me. I never saw him on the river during the winter and it seems he spent that part of the year trading in Wales. The family regarded themselves as being of a much better class than the people we knew in those days as tinkers. I always enjoyed watching Hubert fish his favourite bait, which was a wet fly called the Orange Grouse. In order to attract the trout, Hubert would jig the fly towards him giving his line several short sharp pulls. The part of the activity which really amused me, and which was always a part of the ritual, was Hubert's way of calling 'troutee, troutee, troutee', in time to the jigging line. Strangely enough the technique seemed to work as the bag on Hubert's shoulder was always full of 'troutees'.

It was from the poacher Harvey that I learned the art of fly-tying. He showed me how to make simple patterns of small flies of the kind used generally in County Wicklow. They all had names I could never forget – March Brown, Greenwell's Glory, Bluebody Black Hackle, Red Spinner and the Wicklow Killer. Harvey taught me how to stalk trout before casting a fly on the water. 'Walk very slowly and quietly along the bank and keep your shadow off that part of the river,' he would say, 'Remember trout can feel the vibrations as you walk along the bank.'

Otters have always lived in the Dargle. I often saw their signs and tracks in the mud and sand along the pools where they had been active the previous night. The poacher Harvey often talked about their haunts and their angling abilities. They were really good fishermen, he told me. He also said that they were not really in competition with him, as otters preferred eels whereas salmon and sea trout had far more appeal for him.

The school holidays were never long enough for us as there was always something to do on the river. A big wooden platform arrived on the weir one day. It had been carried down by the current, from somewhere further afield. In the days ahead it was to become a regular means of transport for us in the deeper waters above the weir. The unexplored wooded banks across the river now became the jungles of the Amazon as my brothers and I risked life and limb, dodging crocodiles and hippoes as we poled our craft further and further upstream! Whenever we were not aboard our craft, there were other ways in which we were to learn about life in the river. Building and dismantling dams was a regular pastime and the lifting of rocks and stones revealed the hiding places of an endless variety of aquatic insects and small fishes. There were eels, baby trout, stone loach and lampreys. The harmless loach was known to the local boys as a 'stinger' because of its

Previous page
The large vertical water wheel of Locke's Distillery in Kilbeggan, County Westmeath is still turned by the river Brosna as it journeys from Mullingar close to the summit of the Royal Canal to join the Shannon near the outlet of the Grand Canal at Shannon Harbour.
photograph: Liam Blake

Lowland rivers such as this at Cahir, County Tipperary often have falls or rapids to produce the motive power for mills.
photograph: Liam Blake

set of whiskers! They never referred to the lamprey as anything but a bloodsucker, because of this fish's habit of attaching itself to a stone by holding on to it with its mouth. The young mottled brown herring gulls were called horse gulls, and siskins were known only as devines. My favourite river birds were the kingfisher, because of its exquisite colouring, and the dipper, which amazed us all by regularly walking underwater in its hunt for caddis larvae.

The really exciting days on the Dargle were the flood times. It took two days of heavy rain to turn the Dargle from a small, clear and fairly easy-flowing river, into a mud-coloured roaring torrent. As the water level rose rapidly, it had the effect of making the weir sink from view. The salmon pools then disappeared and the Dargle raced completely out of control, in a mad rush seawards.

For us, it was a good lesson in the power of water. Large trees growing further upstream were torn from their stands along the river bank and came hurrying down like ancient craft, bobbing and dipping as if in salute to their small audience watching from the kitchen window. On one occasion, an unfortunate waterhen chose one of these floating trees as a perch. One could not tell when the unhappy passenger had gone aboard but now it moved uncomfortably from branch to branch, wondering where to disembark: we all cheered as the multi-masted ship passed on downstream with the waterhen still on deck! I often wondered whether the bird eventually did get ashore or if it come to some unhappy end as the tree crashed its way past other obstacles further downstream towards the town of Bray.

At its normal level the Dargle was by no means a large river. In fact a Shannonsider would almost certainly call it a brook. However, the Dargle river's small size was never an indication of its importance as a nursery for young salmon and sea trout. From April to May the whole river seemed to teem with silvery smolt of about four to six inches in length. These were the fish which had changed their troutlike river outfits for sea suits of sparkling silver. They were baby sea trout and salmon answering the call of the Atlantic and working their way downstream towards the estuary. These smolt would spend several years in the sea, even travelling as far north as Greenland in search of rich feeding. Eventually the urge to return would drive the salmon back in the direction of Ireland and instinct would then steer them towards their spawning beds in their native Dargle.

These were the magnificent fish that we would see from our window, jumping, leaping and tumbling as they battled with their first obstacle since exchanging their saltwater habitat for a freshwater one. For these fresh-run salmon and sea trout, the weir was only a temporary barrier. They were as full of energy as the wild Atlantic which had fed them for the past few years. If the rush of water over the fall was too strong, they would try again and again, sometimes moving to another part of the weir to make another mighty leap into the swirling flood waters. It was a wonderful spectacle to watch.

The first migrants to arrive were the large sea trout of three and four pounds, which were the spring fish. They would usually swim into the first flood in June. In the following weeks would come the large runs of salmon weighing from six to twenty pounds and along with them would come in their hundreds the shoals of sea trout weighing anything from three-quarters of a pound to three pounds. These smaller sea trout, which were returning to the Dargle for the first time, were always known in the Bray area as 'clowns'. The name, according to my father, must have been in existence for a very long time, as 'clown' was really the Irish 'caille abhainn', river maiden, which is just what these fresh-run sea trout were!

That is part of what living and growing up in County Wicklow was for me. Little Ned used to say that whenever he was on a river he was close to God, a sentiment which must have often had his angling auidence wondering if God was a little deaf or if the sound of the weir drowned out the punch line in some of Ned's stories! The composer, Seán Ó Riada, who also fished with me on the Dargle, got musical inspiration from several rivers, including his own beloved Sullane in west Cork.

Whatever it is about our rivers and streams, I will be forever grateful that my parents decided to rear their family on the banks of the Dargle river.

Éamon de Buitléar
July 1985

1 How the river flows

Water flows downhill, so naturally rivers run from the mountains to the sea. Since the hills in Ireland lie mostly near the coast there are two main types of river. Those that rise in the seaward side of the hills run quickly to the coast and they are often acidic and rocky. The others, which rise on the landward side, make their way slowly and at times rather aimlessly towards the sea, their richer water flowing past banks of silt and mud. Where there is a depression in the landscape their waters tend to accumulate as a pond or lake until they can overflow to continue their journey. The Shannon lakes lie in such depressions, perhaps scoured out by ice in the Ice Age, but Lough Neagh is an even clearer case. The rivers flow into this basin from all sides and their converging pattern is only broken in the south-east corner where the Lagan flows towards Belfast.

The headwaters – that is the waters near the source – of any river are forever eating back into the hillside enlarging their sphere of influence. Their narrow valleys become broader and they come to take drainage water from a larger and larger area. This process is easy to see on eroding blanket bog or on the seashore where a spring flows out on a sandy beach. By this erosion the stream can capture water that formerly went in different directions. This is what the Lagan seems to have done. Its course lies on soft sandstones and it has worked back to the south-west to divert the flow of the river from Dromara which used to reach Lough Neagh. River capture is quite common: it may be noticed too on a map of the southern part of the Wicklow mountains with the Derreen river.

The Course of the River

In their path to the sea, rivers choose the lowest land even before they start to erode it. It is therefore a bit disconcerting to find that many of our main southerly-flowing rivers have dug through sizeable hills rather than following a seemingly easier route around these obstacles. The Slaney, for example, having risen on the west side of Lugnaquilla, should tend to flow into the Barrow, but instead it turns south-east and ploughs across the mountains in the gap of Bunclody. The Barrow itself has cut a gorge below Graiguenamanagh and the Nore a similar one at Inistiogue. They isolate Brandon Hill in between with its curiously named settlement, The Rower. The Shannon strikes southwards through the Slieve Bernagh hills at Killaloe rather than flowing

out onto the Clare lowlands at Scarriff. The Cork rivers are famous for their idiosyncratic behaviour. Both the Lee and the Blackwater run nice gentle courses eastward from the mountains until they are almost within sight of the sea. Then they turn abruptly south at Cork and Cappoquin and dig their way through three or four separate ridges of rock before releasing their water into the sea.

To find answers to these paradoxes we must go back to the glacial period and beyond. Where the course of a river now cuts through a hill there may have been a time when it flowed at a higher level on a surface now totally vanished. Flowing on this surface it cut its valley into the new rock below, working deeper and deeper while the former supporting rock was weathered away. Alternatively the hills may have risen so slowly that the river has been able to keep pace, cutting its valley through the rising ground. There is nothing strange about land rising in this way: geologically speaking it happens all the time. The raised beach around most of our northern coast and in places elsewhere shows that the land has risen out of the sea by up to 8m in the relatively recent past. Old sea caves occur quite regularly on the landward side of the coast road. Land continues to show a measurable rise in Scandinavia and the Himalayas seem to be adding about a centimetre to their height each year.

There is also the complicating factor of ice to remember. Before the coming of the glaciations the central plain may have been a karstic limestone landscape as the Burren is today. There were hills about but the rock was full of holes and passageways. The Dunmore and Mitchelstown caves are probably relics of this time and there have been many other vertical pipes found that once took surface water down to nether regions. Before glaciation, therefore, our midland river systems may have been very different to what they are today.

The legacy of the two ice advances that took place is seen in physical changes to the hills. There may be similar changes on the lowlands also, but they lie concealed beneath a sheet of rock waste or glacial drift. This drift has an equally great or even greater effect on the rivers. For one thing it allows them to flow on the surface again. In a limestone region such as the Burren, where drift is almost absent, rivers flow largely underground, though they sometimes briefly visit the surface. The Fergus, for example, disappears eight or nine times on its journey to Ennis, and its tributaries spend practically all their time hidden from view. Glacial drift is also piled into moraines, drumlins and eskers. In a flat lowland these deposits may control the course of the river. The Shannon in fact has eroded no valley in the middle part of its course. Its gradient (slope) is too flat and its flow too slow to mould the landscape. It contrives to flow south as a sort of moving lake, hemmed in by nothing but soft glacial deposits and bog. The absence of meanders (bends) suggests that it is no normal river.

Drift can have more dramatic effects: it may block a former valley and force a river to change its course. The Liffey in Kildare offers an example, for it used to flow westward from the Wicklow mountains to the Barrow. After the retreat of the ice it found the Curragh deposit of sand and gravel directly in its way and had to cut a new course through Kilcullen and then on to Leixlip. The Vartry river in Wicklow was blocked at Roundwood and was diverted south-east, where it is still cutting a fine valley, the Devil's Glen. Similarly the Bush river in Antrim had to abandon its former course down to Ballycastle in favour of a longer and less visually attractive route to Bushmills. Glacial excavation is common where the ice was confined by hills, and several modern rivers follow the path of the glaciers. The Newry river flows down an ice-cut channel into an ice-cut fjord in the shape of Carlingford Lough. The Slaney follows such a route at Ferrycarrig near Wexford while the Erriff's course in Mayo has been so modified by ice erosion that its flow may actually have been reversed. The Shannon at Killaloe is subject to two glacial modifications. Here the ice seems to have branched the Slieve Bernagh ridge in the same way as it took the bite out of the Devilsbit mountain nearby. It also left enough debris at Scarriff, including drumlins, to block the former outlet of the river. Behind this block the water now flows back to Lough Derg in the Scarriff river. The main Shannon has to head southward through its new gap and it plunges – or plunged before the power station was built – over the falls at Ardnacrusha as if in relief. The presence of a waterfall such as this is a clue to a fairly recent change in the river's behaviour, as on a longer timescale it disappears into rapids or riffles.

When the ice melted it released a great quantity of melt water capable of cutting sizeable gorges. The Glen of the Downs in Wicklow illustrates its potential for erosion. All of our southern rivers and later many of the others must have taken unimaginable volumes of water to sea at this time, for the ice had been up to 1000m thick. It seems likely that it caused one other peculiarity of these rivers, the fact that the sea comes a long way inland up their valleys. Sea level was lowered during the Ice Age when so much water was frozen into the ice mass. It did not quickly return to normal and the rivers cut deep trenches down to it, burgeoning with their debris-filled flood waters. The rise of the sea eventually drowned the lower parts of the valleys, partially filling them with sediment, but allowing the navigation by ships that today is such a feature. Elsewhere too melt waters were important, and it is thought

The abundant Irish rainfall gives rise to a dense network of streams and rivers which can be seen on this map on the western side of the country. This network was added to by the eighteenth-century canal system seen here on the eastern side of the country.

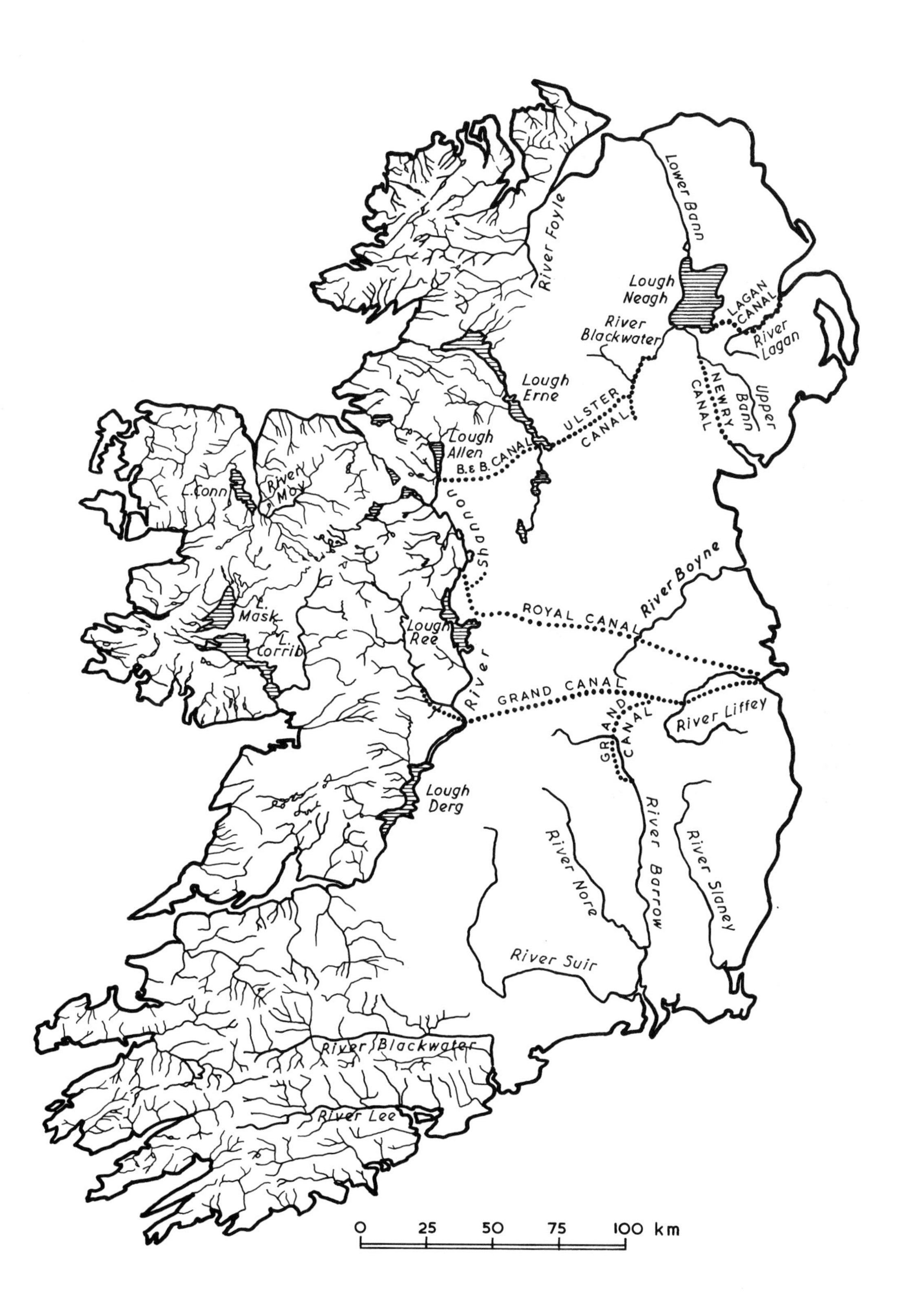

Lower Bann
River Foyle
Lough Neagh
LAGAN CANAL
River Lagan
River Blackwater
NEWRY CANAL
Upper Bann
Lough Erne
ULSTER CANAL
Lough Allen
B. & B. CANAL
L. Conn
River Moy
Shannon
River Boyne
L. Mask
L. Corrib
ROYAL CANAL
Lough Ree
River
GRAND CANAL
GRAND CANAL
River Liffey
Lough Derg
River Nore
River Barrow
River Slaney
River Suir
River Blackwater
River Lee
0
25
50
75
100 km

that the Owenmore finds a glacial spillway through the arc of the Mayo hills at Bangor.

The Flow of the River

Where do rivers come from? We have already had one answer to this question – they come from the mountains and they move towards the sea, eating out paths for themselves or following ancient channels left by the Ice Age. But where does the water that makes up the river come from?

Most ancient peoples thought springs and rivers came from a great lake below the surface of the ground. Both the Greeks and the Romans wrote of a series of underground channels connected to an underground ocean. In the first century AD Pliny wrote in his *Natural History:*

> Water penetrates the earth everywhere, inside, outside, above, below, along connecting veins running in all directions. . . it breaks through to the highest mountain summits and there it gushes as in siphons.

This belief is understandable among Mediterranean peoples who lived in lands that have little rainfall through much of the year and where there is a limestone or karstic landscape with numerous caves and underground streams. But this belief seems to have been widespread in Ireland also, although rainfall was abundant. In Celtic Ireland lakes, rivers and springs were considered to be a source of contact with the well of the Otherworld which lay in the centre of the earth and was the source of all wisdom and knowledge. To defy the magic powers of this well was to commit a serious offence. One legend describes the breaking of three huge waves on the goddess Boand in retribution for such a misdeed. The goddess fled to the sea and the waters followed her, and so the river Boyne came to be. There is a similar story about another goddess, Sinann, for the origin of the Shannon.

The Book of Invasions or Lebor Gabala deals with the origins of physical features and of names. It relates that before the invasions of Partholon there were only three lakes and nine rivers in Ireland. After this time fourteen more lakes appeared, most of them being associated with the digging of an important grave. Such an increase in the number of rivers and lakes is not in fact incompatible with the climate of the time, for the Ireland of 1000 BC was wetter and colder than that of 5000 BC. The expansion of peat on the hills that replaced the former forest cover took place around 2000 BC, and this was accompanied by an increase in the size and number of rivers and lakes as reflected in

The clean acid waters of the upper Liffey here at the Sally Gap in County Wicklow are home to a wide range of fly larvae and nymphs. The adults hatch at different times between March and October.
photograph: Richard Mills

folk memory.

The picture of underground water as the source of springs and hence of rivers persisted through classical times, the Middle Ages, the Renaissance and almost to the present day. Leonardo da Vinci wrote in AD 1500.

'rom *The literary works of Leonardo Da Vinci* edited and compiled by Jean Paul Richter OUP 1939

> The waters circulate with a constant motion from the utmost depth of the sea to the highest summits of the mountains, not obeying the nature of heavy water. In this case they act as does the blood of animals which is always moving from the sea of the heart and flows to the top of their heads. . . When the water rushes out of a burst vein in the earth, it obeys the nature of other things heavier than air, when it always seeks the lowest places.

An upland stone-filled river caught in the act of short-circuiting an old meander (on the left). It is digging into the new bank moving material downstream.
photograph: Richard Mills

It is only in recent centuries that the notion of water coming from within the earth has been displaced by the idea that *precipitation* (rainfall, snow, dew etc) is the source of flowing water. However, there is one suggestion of this concept in early Irish literature in the story which makes Uisneach, west of the present Mullingar, the fount and origin of our river system. This account tells how there was a great hailstorm at Uisneach on the occasion of the inauguration of Diarmait son of Cerball as king. Such was its greatness, it is said, that the one shower left twelve chief streams in Ireland for ever. A later version of the story attributes the twelve rivers radiating from Uisneach to St Ciarán, who worked a miracle to break a prolonged drought.

We now know that the flow of rivers occurs because *precipitation* exceeds what is lost from the land surface by *evaporation*. In other words, more water falls on the earth than is soaked up again by the heat of the sun. The excess *runs off* to the ocean either rapidly on the surface or more slowly through the ground.

Over the ocean, on the contrary, evaporation is greater than precipitation, so the air mass over the sea has a high humidity. This moisture-laden air is carried in over land by wind, the water vapour condenses into clouds and eventually falls as rain or another form of precipitation. The sequence of evaporation, precipitation and run-off is called the *hydrological cycle*.

Whether precipitation is enough to sustain all river flows was for long a subject of debate among scientists. This dispute can only be resolved by the measurement of all elements in the water balance and this has not been done over large areas of the continents or the oceans. Most modern estimates put the average precipitation over land at about 850mm per year, the overall evaporation from soil and plants at 550mm, and the resulting run-off which feeds the rivers at 300mm.

Mosses and algae coat the larger stones in a fluctuating mountain stream where dampness is constant.
photograph: Richard Mills

The people who made significant contributions to the modern concept of the hydrological cycle are better known for their work in other fields. Bernard Palissy published the first clear exposition of the argument for rainfall as the origin of run-off (rivers) in 1580. He is better known as an innovative potter, a pioneer geologist and a persecuted Hugenot. Pierre Perrault, a former superintendent of taxes, measured and compared rainfall and river flow in the upper part of the Seine catchment in 1674. Though his methods were crude and his arithmetic faulty, he reached the correct conclusion, that the annual rainfall exceeded annual stream flow and that no supply of water from under the ground was necessary to sustain the rivers.

Edmund Halley, the astronomer, worked out the evaporation of water from the Mediterranean in 1687 and the inflow of rivers to it. He concluded that the flow of the rivers would be just over one-third of the evaporation and though this figure is none too accurate, evaporation does account for the inflow of water through the Straits of Gibraltar. The first person to estimate rainfall, river flow and evaporation separately for a given area was John Dalton, who in 1799 presented a paper to the Literary and Philosophical Society of Manchester. It was entitled 'Experiments and observations to determine whether a quantity of rain and dew is equal to the quantity of water carried off by the river and raised by evaporation: with an enquiry into the origin of springs'. (Dalton is more remembered of course for his contribution to atomic theory.)

Our Irish climate is affected by the expanse of the Atlantic Ocean on the western side of Ireland and the prevalence of westerly and south-westerly winds over it which drive warm water towards us in the Gulf Stream and the North Atlantic Drift. Precipitation is usually in the form of rain or drizzle rather than hail or snow, dew or hoar frost. If rainfall and evaporation were both evenly distributed throughout the year then the flow of Irish streams and rivers would be uniform throughout the year, but in fact the flow in winter is distinctly greater than the flow in summer. This is due not so much to differences in rainfall between winter and summer, but to higher levels of evaporation in the summer months.

Rainfall in the Shannon catchment is almost equally divided between winter and summer, and this is a good example to show the importance of evaporation to the seasonal variation in run-off or river flow. The average annual rainfall on the Shannon catchment is about 1010 mm. The average long-term outflow from Lough Derg, about 175 cubic metres per second, is equivalent to a depth of about 470mm on the whole catchment, so that the run-off is just under half the annual rainfall of 1010 mm. Unlike the rainfall, however, the run-off is quite different in summer and winter. The average run-off for the six months from October to March is about 259 cubic metres per second, compared with an average of 97 cubic metres per second for the six months from April to September. Since the rainfall in the two halves of the year is almost the same, the difference in run-off is almost all due to the difference between winter evaporation of 90mm and summer evaporation of 450mm.

As this example of the Shannon shows, the amount of precipitation on any closed catchment area – a catchment area is a basin of land where rain water collects – will equal the sum of evaporation and run-off, in the long term. For shorter periods, however, this balance has to be adjusted to include changes in the storage of water within the catchment. Water can be stored on the surface of the earth in lakes or in river channels, or else below the ground. It accumulates below the surface in the upper soil layers, but below this again there is a saturated zone, called the *groundwater*, from which springs take their supply. The boundary between the layer of damp soil and the groundwater is the *water table*. The depth of the water table below ground fluctuates over the year.

As we have seen, when precipitation exceeds evaporation, as it does in Ireland in the winter months, the flow in the rivers increases. However, part of the excess precipitation will be used to increase the amount of water stored within the catchment area. The part of the precipitation that does not immediately go to swell the rivers infiltrates downwards through the ground surface to replenish any deficiency in soil moisture and any excess will recharge the groundwater storage. From May to July evaporation exceeds precipitation, but the rivers continue to flow in these months by drawing on that accumulated water storage and in particular on the lakes and groundwater. In August and September precipitation may again exceed evaporation, but the excess precipitation will go to fill up the depleted underground storage, so increased run-off (in other words, fuller rivers) does not usually occur until October or later. If the rainfall in the summer months is less than average then the low flows will continue into the autumn. Most streams and all rivers in Ireland continue to flow throughout the year and the catchment storage is replenished in late autumn and winter. However, prolonged periods of low precipitation and high evaporation – drought in other words – can produce not only low flows but also such a depletion of natural water storage that it may not be made up again in one winter season.

Droughts in Irish History

The earliest recorded meteorological events in these islands are to be found in the Irish annals. The exact nature of a phenomenon and the exact date of its occurrence are not always easy to deduce. A four-line poem from the Annals of Ulster describes three showers falling at Ard-Uilline at the end of a drought:

> Three showers at Ard-Uilline
> fell through God's love from heaven:
> a shower of silver, a shower of wheat,
> and a shower of honey.

We would attach no great hydrological or meteorological significance to this verse except for the fact that we have a fuller reference in the Annals of Clonmacnois. The original of the latter manuscript is now lost but the text is known to us through a sixteenth-century English translation:

> There was a Great famine throughout all the kingdome in the beginning of his raigne, In soe much that the King himself has very little to live upon, and bein then accompanied with seven godly Bishops, fell upon their knees, where the King very pitifully before them all besaught god of his Infinite and Great Mercy if his wrought otherwise could not be appeased, Before he saw the Destruction of so many thousands of his subjects and friends that were helpless of a releefe and Reday to Perrish, to take him to himself, otherwise to send him and them some Releefe for maintenance of his service, which request wasnoe sooner made than a Great shower of Silver fell from heaven, whereat the king Greatly Rejoyced and yett (said he) This is not the thing that can Deliuer us from this famine and eminent Danger, with that he fell to his Prayers againe; and then a second of heavenly honey fell, and then the king said with Great thanksgiving as before, with that ye third shower fell of pure wheat, which covered all the fields over that like was never seen before, soe that there was such plenty and aboundance of wheat, that it was thought yet it was able to maintaine many kingdomes. Then the King and the seven Bushopes gave great thanks to the Lord.

Of course we cannot be sure that these two accounts refer to an actual drought and ensuing famine; they may be merely a poetical account attempting to explain the origin of the name of Niall Frosach (Niall of the Showers) who was later High King.

It is difficult to date such events and to co-ordinate them with accounts from other countries – the verse in the Annals of Ulster attributes the above event to AD 763, whereas the Annals of Clonmacnois assign it to AD 759. What is clear from the entries in the annals is that the eighth century was one of recurring drought in Ireland in contrast to the preceding and succeeding centuries. Thus between AD 713 and AD 775, no fewer than fifteen severe droughts are recorded by the Irish and British annalists. In contrast only one notable drought is recorded in the ninth century and even that drought in AD 822–823 was noted as being most severe in England. The next notable series of droughts was between AD 987 and AD 994, when five of the eight years were recorded as particularly dry.

In later centuries the entries in the annals are less poetic in form but are still interesting. The Annals of Innisfallen record for 1129 that there was a torrid summer in which the streams of Ireland were dried up. The Annals of the Four Masters and the Annals of Lough Key both record the drying up of the Galway river in the last quarter of the twelfth century but attribute the event to different years. The Four Masters record that in 1178

> the river of Galway was dried up for several days, so that all things lost in it from time immemorial were recovered, and great quantities of fish were taken by the inhabitants.

The Annals of Lough Key record for 1191 that 'the Gallimh became dry this year'. There were apparently two successive years of severe drought in Ireland and England in 1325 and 1326. It is recorded of 1326 that

> in this year there was an unprecedented drought in Ireland: so that in Winter there was not much rain and in Spring, Summer and Autumn, there was none, and very great dryness and great heat, so that the brooks and great rivers (which always gave plenty of water) were almost dry.

Sir William Wilde included a table of 'Droughts and Heats, Hot Summers and Mild Winters' in part V of the Census of Ireland for 1851. This is the volume which contains his famous report and tables concerning pestilences and causes of death. From this table of droughts, based on descriptive accounts and early instrumentation, one can identify extended droughts lasting for more than one season in 1736–37, 1740–41 and 1760–62. More detailed information on these later droughts is available from extant diaries and from newspapers.

With the nineteenth century we have the beginning of reliable continuous instrumental records which allow us to identify on the basis of rainfall records dry years such as 1887, 1921 and 1934. Direct flow measurement of the river Barrow in 1934 gave a run-off of 1.1 cubic metres per second from an area of 1660 square kilometres. This is equivalent to 0.7 litres per second per square kilometre or roughly one drop per day per square metre.By 1959, which was another dry year, a large number of hydrometric stations had been installed and rated. Consequently, low flows during this and subsequent years could be directly measured or inferred from the recorded water levels. These results indicated that in a year such as 1959 the low flow of Irish rivers was normally 0.5–2.0 litres per second per square kilometre of catchment. 1984 was another exceptionally dry year and the low flows in the Shannon region in that year approached the lowest recorded flows.

Floods in Irish History

Information concerning flood flows in rivers can also be derived from literary sources. The first records of floods on the Shannon are from the tenth century. The Four Masters record that in AD 920 there were

> great floods and the waters reached the Abbey of Clonmacnois and to the road of the three crosses in Ulster.

Again in 942 there was flooding and half of the eastern part of Clonmacnois was demolished.

The Annals of Lough Key record 'prodigous tempests and great moisture' for the year 1037. Since other annals have similar reference there must have been widespread flooding in that year. Other references to floods occur in the Irish annals for 1050 and for 1094.

Following the Norman invasion in 1169 we find references to Irish weather conditions in the English chronicles. Thus Benedict of Peterborough in his description of the visit of Henry II to Ireland in October 1171 says:

> And the King of England could by no means finish the war during the Winter, by reason of the overflowing of the waters, and the steepness of the mountains.

Another Shannon flood is described in the Annals of Lough Key for 29 June 1251, which records that a great shower fell on the festival day of Peter and Paul so that a boat sailed all around the town at Kilmore on the Shannon.

In the succeeding centuries there are more frequent accounts of the effect of floods in urban areas. The following is a reference to the floods in Kilkenny on 17 November 1338:

> Also on Tuesday, the 15th of the Kalends of December, there was a very great inundation of water, the like of which had not been seen for forty years previously in which bridges, mills and buildings were overthrown and carried away; the waters did not reach to cover the foot of the great

altar of the steps of the altar of the whole abbey of the Friars Minor at Kilkenny.

Later records indicate that in 1705, half of Limerick was drowned and in 1742 there was a great flood and inundation in Limerick. The conjunction of floods and droughts in the same year or in succeeding years is also evident from the historical records. The drought of 1761–62 was apparently broken in October 1762 when the North Wall in Dublin was overtopped and many bridges were broken down. Floods are reported in many parts of the country in the succeeding years 1763, 1764 and 1765. Damaging floods are recorded for Cork in the years 1789, 1853 and 1916. The level of run-off in the 1916 flood has been exceeded in recent years, notably in 1978, but the effect has been modified by Carrigadrohid and Inniscarra reservoirs.

Measuring River Flow

In the nineteenth century measurements of rainfall and run-off became increasingly important as the basis for the engineering design of new water supplies and of arterial drainage schemes. Robert Manning described in an 1866 paper how he set out to design a new water supply for the city of Belfast. Manning erected three rain-gauges and two measurement weirs in the Woodburn catchment near Carrickfergus and observed them for twelve months. He then extended this one-year record of rainfall by comparison with the fourteen-year record at Queen's College Belfast, 14km away. By comparing the rainfall and river flow in the Lisburn catchment, he was able to estimate the variation of monthly residual of rainfall minus run-off throughout the year. Using his estimate of these 'losses' (vary from 5mm per month between November and January to 70mm per month between August and October), Manning was able to estimate the monthly supply of water over a fourteen-year period.

In early design of arterial drainage works, the river flow to be catered for during floods was taken to be a proportion, usually between one-third and two-thirds, of the maximum recorded daily rainfall in the nearest rain-gauge. William Mulvany, who was Commissioner for Drainage from 1842 to 1853, encouraged his district engineers to take every opportunity of checking their designs against measured flows. The flows on the smaller streams were measured by means of weir-gauging and the flows on the larger streams by float measurements. The years between 1842 and 1850 were relatively free from floods but remarkably high flood flows occurred in 1851 and 1852. During the expansion of the drainage works in the famine years there were at times engineers resident in over a hundred districts and much valuable data was obtained. The results of these flow measurements were returned to the head office of the Board of Works and tabulated as a basis for the design of new works. The results indicated quite clearly that the flood run-off per unit area was higher in smaller catchments than in larger ones.

What is called the rational method for the estimation of flood peaks was developed in Ireland during these famine years. This approach is based on the assumption that the maximum flood run-off will be generated by a storm which lasts for the time taken for the rainfall to travel from the most remote part of the catchment to the outlet. This concept of the time of concentration was formulated by Thomas Mulvany, District Engineer in County Cavan, in a paper to the Institution of Civil Engineers of Ireland in 1851. He had installed a continuously recording rain-gauge and a continuously recording flow recorder to show when maximum flow was attained. His work is directly quoted in hydrology lectures today.

In the first quarter of the twentieth century direct flow measurements were taken in connection with proposals for hydroelectric installations on the Liffey and the Shannon. The Electricity Supply Board has continued to carry out river gauging on those rivers in Ireland with potential for hydroelectric power and maintains records for sixty stations. Since 1939 the Office of Public Works (formerly the Board of Works) has operated a systematic and comprehensive hydrometric survey. Since the early 1950s most of their gauging stations have been equipped with continuous water-level recorders. This hydrometric network, whose main purpose is to provide data for the design of arterial drainage schemes, has resulted in an archive of continuous records for about 170 gauges, many of which extend back over twenty-five years or longer. An Foras Forbartha (the National Planning Institute) has been active on hydrometric work since the early 1970s and now maintains 150 continuous recording stations. All three agencies process their hydrometric data and compile it in useful forms. They make it available to interested people.

Overleaf
The marsh marigold flowe[rs] early before grass or other vegetation is properly grown. It grows in muddy places out of reach of most floods.
photograph: Liam Blake

A lowland acid river enters [a] lake in Connemara, Count[y] Galway. This is the way all sea trout and salmon must pass in their search for spawning beds upstream.
photograph: Liam Blake

2 A watery world

The stems of water crowfoot flow with the current and in this way protect themselves from damage. The plant flowers in early summer whenever levels are at their lowest.
photograph: Richard Mills

Previous page
The yellow flag is an iris with three big petals for its flower. Each year the leaves spring up from a stout root stock which stores food.
photograph: Liam Blake

The square stemmed purple loosestrife is a common plant of wet meadows and river banks. Its dead stems persist long after the flowers have faded.
photograph: Richard Mills

Look over any bridge in Ireland and, apart from the refuse of our civilisation, you will see a different world, one of waving water weeds and skimming flies. The plants and animals below you live in water, not in air, but they still need oxygen to breathe and light to grow or food to eat. They must withstand a constant pull downstream by the water's flow, the scouring and buffeting of floods, the smothering effect of silt and also all those waste materials that we let into their habitat. But they are protected from extremes of weather, from frost, heat and drought. The watery world is a sheltered environment in which life first evolved. Fresh water was the route by which early marine organisms colonised the land giving rise ultimately to forest trees, birds, insects and mammals. Curiously enough, most of our larger water plants and some of the animals too, like snails and beetles, are really land organisms that have gone back into the water. Only the algae remain in anything like their original form, having changed little in hundreds of millions of years.

The algae or 'freshwater seaweeds' are everywhere and form the basis of life in many waters. They are generally small, even microscopic, but they also form threads or cushions of green. The brown algae of the seashore are almost absent from fresh water, but there are a few reddish ones which you may meet. Algae are adaptable plants: there are species ready to exploit every habitat within the river and they appear and disappear rapidly in response to changing conditions. There are some that encrust rocks with a skin of hard lime so that they will not be damaged, some that move upwards through layers of accumulating mud so that they are never covered and others that form a weft of strands between water plants so quickly that it seems by magic. There are few floating forms of algae in our rivers in contrast to our lakes. Where lakes or backwaters flow into a river, however, some of these planktonic forms will be found. The time they spend in the river before it reaches the sea is not sufficient to allow them to multiply, so their numbers have to be continually reinforced.

The flow of the river has obvious drawbacks if it sweeps all one's offspring away. In fact there is a constant drift of organisms downstream and to counteract this loss many animals will move only against a flow of water and some travel upstream before they lay their eggs. The danger of dislodgement is a very real one in the faster stretches for animals that do not swim well: a slight careless movement may give the current enough purchase to whip off an animal without any hope of reattachment. To get over this many anchor themselves securely. The limpet presses its shell tightly against the rock, the young blackfly spins threads of silken glue to hang on to and the young caddis resorts to ballast and a heavy case of stones. All around the surface of rocks and other objects in even the fastest flow there sits a thin layer of still water. Tiny organisms or flattened larger ones may live in this boundary layer if they can find food here. Mayflies and flatworms commonly do so and some others lay their eggs in the security it offers. The larger animals must seek out quiet areas between stones in which to live. The largest, the fish and the otters, can match all but the fastest flows with muscle power, though they spend most of their lives in much quieter pools.

The speed of a river depends on the slope of its bed and the volume of water coming down. Water flow seems fast in a rushing mountain stream but it is held back by all the turbulence and rocks and, despite the noise, it seldom can compete with the smooth surging flood of a lowland river. You have to run to keep up with the central flow of water, though the margins are slowed by bankside eddies. Great branches are carried down by the floods, half sunken in the water like crocodiles, and rocks are rolled along the bottom, churning up the beds of sand and gravel left during more tranquil times.

Rivers dig out their beds both downwards and sideways. In the hills they have more scope to cut down, and the water with its load of loose stones scrapes out a notch that becomes a narrow valley. Time will broaden this valley and it will work back into the mountain. The stream will become less confined by rock and it will start to meander. When its course levels out it can no longer cut downwards and its sideways erosion will come to predominate.

The meanders work across the flat flood plain of a lowland river changing the course of the river and leaving marshy backwaters and loops in their wake. Willow trees are common on the edges of such rivers. The undercutting of the banks loosens their roots causing them to trail in the water and finally to be washed away.

Flow in a river has its positive side too. It sorts and deposits silt or gravel into banks, making new habitats at every turn, and it brings silt and nutrients downstream where they have an enriching effect on river life and, when the river floods, on the meadows in the flood plain. It also brings a constant supply of water into contact with plants and animals from which they can extract nutrients or food. Many small animals feed by trapping tiny particles from this supply. They spin nets or use a filtering gill for the purpose. Flow also helps in the transfer of oxygen from air to water, as broken water takes it up much more rapidly than still water. Oxygen is also released by submerged plants in day time. You can often see bubbles of oxygen caught amongst strands of floating algae. Animals may take special precautions to get enough oxygen. Some swim up to the surface to trap a bubble of air, others have a long tube or snorkel reaching to the top or into air spaces in plants. Yet others have a special red blood, like our own, to trap the little oxygen there is in the places they live. The roots of plants may have difficulty too, and spongy air-filled stems or leaves down which oxygen may pass are common.

Another benefit of moving water is that it carries away dead materials, which if they were to decompose in one place could use up all the oxygen in the water and thereby kill most of the animals. This can happen in ponds at leaf-fall, but in flowing water the leaves represent a glut of food. Many types of animal chew on fallen leaves or dead wood directly, others eat the bacteria and fungi which colonise them. Looked at in overall terms, at least half of the animal life may depend for food on this input of *detritus* from the outside world.

The chemicals dissolved in a river come from the rocks and soil over which it flows. Peaty soils and blanket bog release few minerals, so the brownish water that flows from the hills is nutritionally poor. In particular there is no lime in the water. Lime-free water is said to be 'soft' and it produces no deposits in a kettle. It is also normally slightly acid extent and it can support only relatively small numbers of plants and animals, of different sorts than those in richer streams. It can be said to be *oligotrophic,* or poor in nutrients. The Caragh in County Kerry and the Avonbeg in County Wicklow are examples of oligotrophic rivers. Just as soft, acid, peat-stained water and hills generally go together, so do the opposites, hard, alkaline, clear water and lowlands. In Ireland not all lowland rivers are hard, but they are if they flow through limestone land, either where the rock lies close to the surface or where limy glacial soil has been left by the ice sheets. The water loses its acidity, it picks up lime and becomes hard and calcareous. The lime and other nutrients make it *eutrophic,* able to support large numbers of organisms. If the flow is slow enough, water-lilies or crowfoot can spread over the entire river, creating an underwater jungle in which survival may be chancy among the snapping jaws of water beetles and the voracious shoals of fish. Coarse fish like bream and pike now share the habitat with trout and eels and create a food source for otters, mink and herons. The Shannon and its tributaries and the Maigue, the Suir, the Tolka and the Camac are all eutrophic rivers.

1. canary grass
2. damsel fly
3. bur reed
4. mayfly swarm
5. bream
6. Canadian pondweed
7. heron
8. bur reed floating leaves
9. water boatman
10. crayfish
11. Cladophora alga
12. water starwort
13. clubrush floating leaves
14. clubrush
15. swallow
16. rudd
17. pondweed
18. shrimp

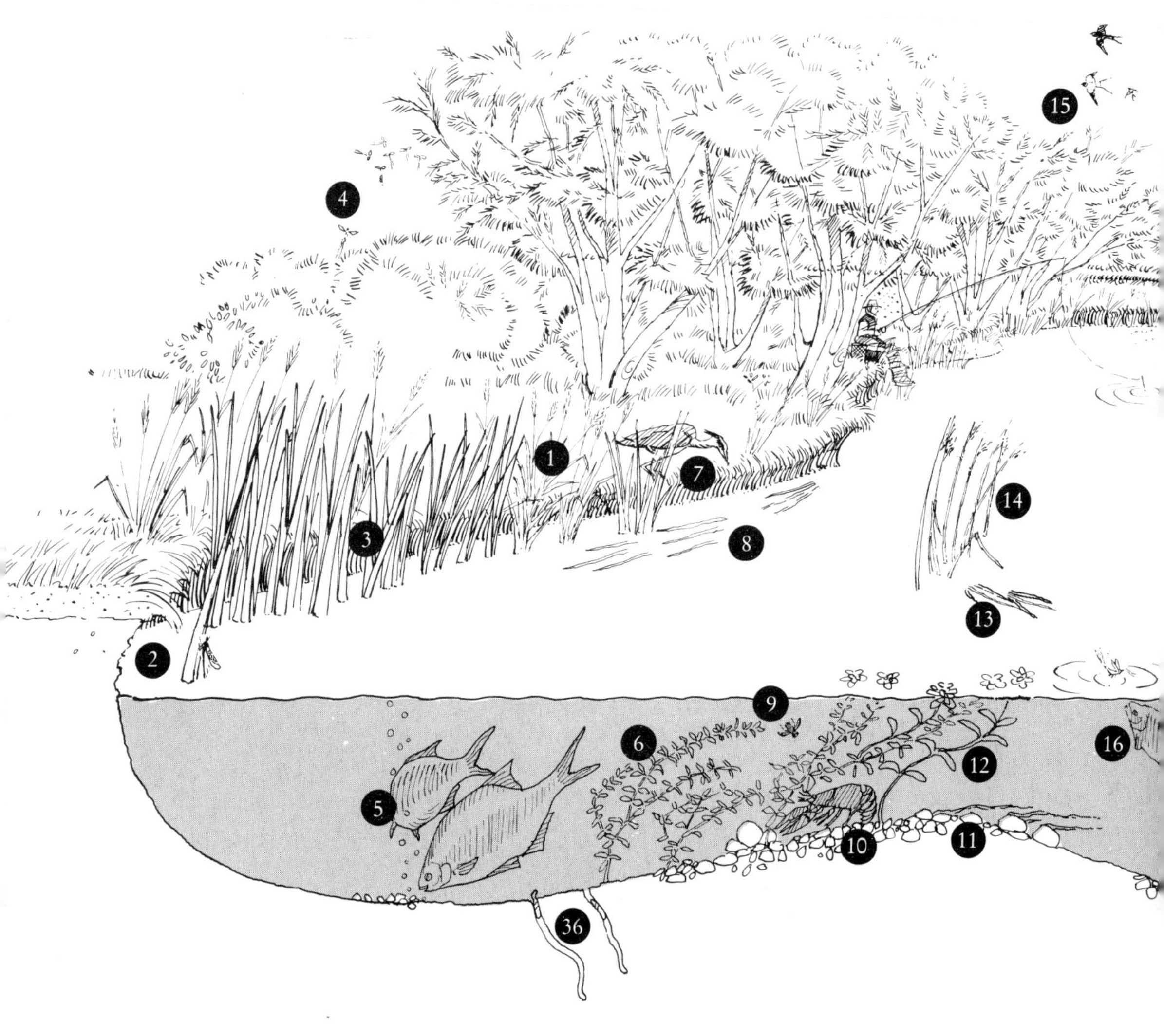

19. ram's horn snail
20. diving beetle
21. water mint
22. mink
23. watercress
24. tench
25. mute swan
26. pike
27. mayfly nymph
28. perch
29. pond snail
30. minnow
31. caddis swarm
32. beetle larva
33. fools watercress
34. yellow waterlily
35. leech
36. mayfly nymph (*E.danica*)

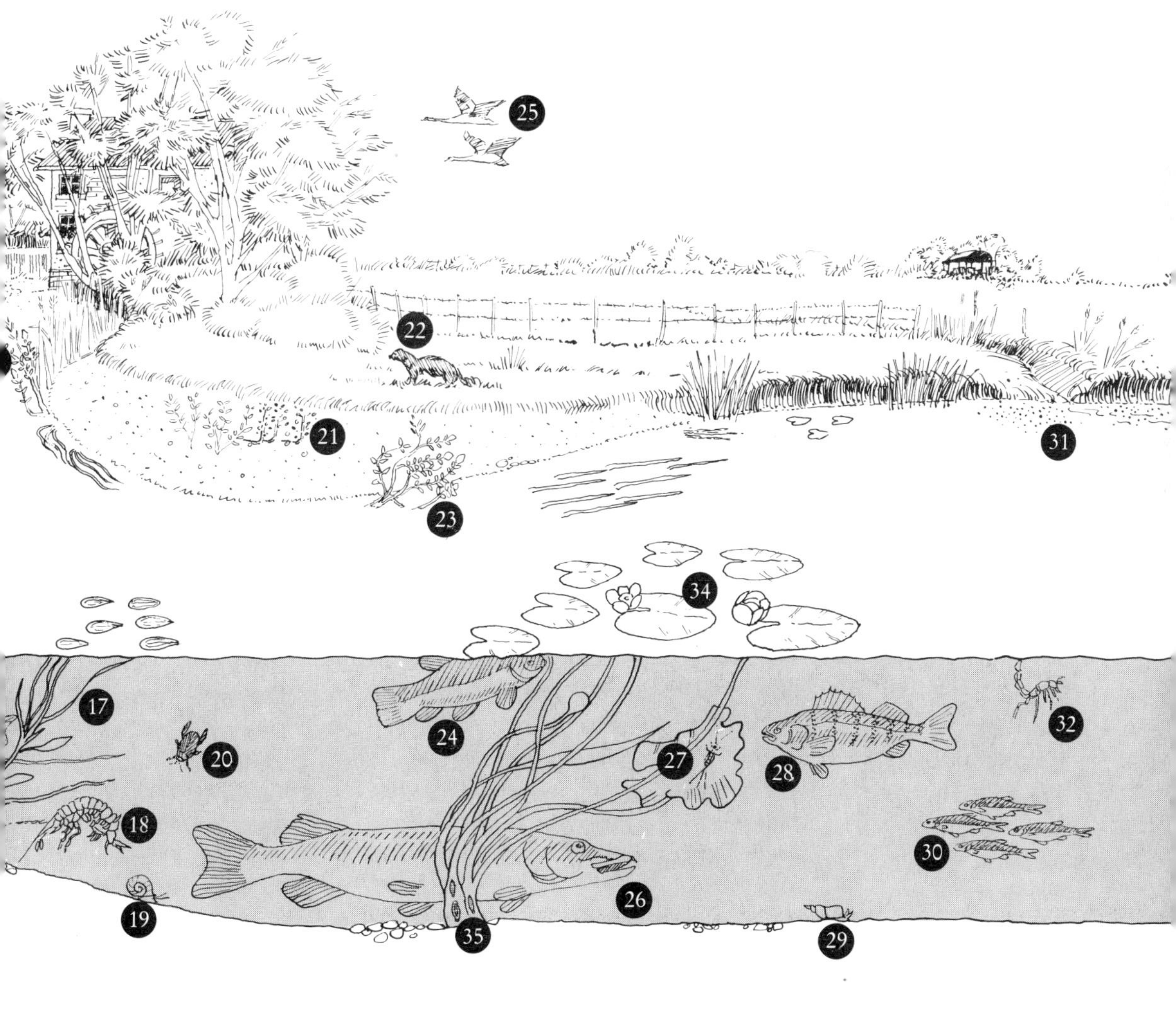

Rivers change over the year. The clear, plant-filled pools of summer picnics are impossible to find in the churning floods of November or January. For the plants spring and summer are the busy times. Floods are rare, though they may occur at any time, and growing temperatures are good. The algae may take advantage of the early spring when light is strong and shade not yet developed. Their sudden growth provides food for tadpoles, snails and the young insects which may require this extra ration for their transition to adult flying forms. The blackflies, mayflies, midges and caddiscs all hatch out from March to October and their dancing swarms feed many a swallow, a sand-martin or a bat before they manage to lay eggs for the next generation. The fish lie in deep pools in the summer, feeding leisurely on surface flies or bottom food and waiting for the autumn rains to awake their migratory instincts and give them access to their spawning beds.

In looking at rivers we are going to start in the hills and work downstream to the sea. Rivers may have simple beginnings, but as they grow larger they become richer in life, more powerful and more complex. They begin to flow among fields, farms and towns and they may be dredged or drained or diverted. The rock type they flow on is very important to their character, so we will note whether it is acid, like granite, sandstone and slate, or calcareous (alkaline), like limestone and chalk.

3 Mountain rivers

Finding the source of a river is never easy and it will test both the stamina and the detective powers of the explorer. Seldom will you find the water springing out of the ground and setting off on its journey to the sea. Even the Shannon Pot, a pool on the side of Cuilcagh mountain in Cavan which is the traditional source of this river, is filled from a substantial area of ground higher up the mountain where the true source must lie. You have to follow the tiny channels across the low boggy slopes of the hills into a maze of seepages, soaks and flushes where only a change in the vegetation may tell of moving water. A patch of rushes or moor grass set in a barely perceptible hollow may be your final destination.

Bogland Beginnings

Peat is the common cover of the hills. After rain, pools and trickles of water are everywhere and each step makes a splash of water. The water soaks into great cushions of moss, it runs together around tussocks of deer sedge and between the stems of bog cotton. It flows over the surface on a thin skin of the pinkish alga *Zygogonium*,* which protects the peat below. If it breaks through this skin it may dig out a tunnel in the peat, disappearing temporarily below ground and picking up tiny particles which produce the brown colour so characteristic of mountain rivers. In many areas burning and grazing have so weakened the bog vegetation that these headwater streams now spend most of their time digging out peat and do much more erosion than at any time in the past. You only have to walk over the summits and saddles of the Mourne or the Wicklow mountains to see that all is not well with the hills. Desolate gullies of loose peat extend their tentacles ever upwards into the intact bog. Bog flows and slides widen the channels, eventually revealing the rock beneath and leaving isolated peat hags in between, a sad reminder of the past. As mentioned earlier, this is an excellent place to see stream erosion in action, mimicking on peat what all our rivers have been doing to soil and rock over thousands of years.

Run-off from intact bogland is more regular than from many other habitats because the peat acts as a sponge, soaking up rain when it can and releasing it gradually. Peat may be 95 per cent water when it is

* All organisms have two Latin names, the name of its genus and the name of the species. In this book the generic name is used on its own, unless this would be confusing.

fully wet, so it has quite a capacity for water-storage. It does get full up after rain, however, and in this case a further shower will run off very fast and may create a flash flood downstream. Short steep rivers like the Dargle or the Dodder, which flow seawards from the Wicklow mountains, are like this and have caused notable floods in the past. A longer, flatter river can overflow its banks in places and the valley can then store some of the water until it can flow back into the channel. Floods further the work of erosion as they shift loosened material down towards the sea as well as cutting into the river banks. During dry periods the bog streams are surprisingly clear and their tinkling and gurgling add to the other sounds of the mountain, the wind, the sheep and the meadow pipits.

Peat is formed of undecomposed plants and is covered by a skin of living vegetation. The rain that drains through it is deprived of its few dissolved nutrients and is given weak humic acids in return. This water is therefore a very poor source for plant and animal growth. As soon as it burrows its way down to the rock below it can pick up minerals to add to its chemical load. The granite and gneiss of Donegal or Tyrone are themselves nutrient-poor, but the slates of south Mayo and the basalts of Antrim are somewhat richer and this is reflected in the variety and amount of animal life in their streams.

Mountain Stream Habitats

Mountain streams start out with few plants or animals. However, algae of a rather special type do grow on the exposed rocks and sometimes on the peat. Nitrate and phosphate, which are essential for any kind of life, are present in rain water, and so in bog waters, at only extremely low levels. Some of the algae get over half of this problem by trapping nitrogen from the air or from the water and turning it into useful compounds. They are the blue-green algae, among the oldest and most primitive of plants. Some give sustenance to a few tiny animals too, especially the young stages of midges that in early summer change into adults and become more bloodthirsty.

Rapids and pools, shallows and deeps alternate in streams in a way that produces many different habitats. Thus at the edges even of a fast stream there may be pond skaters which require quite still water. You will see them skimming out on the surface film to seize a floating insect and then retreating under the bank to eat it. Overhanging banks, often clothed in the green straps of the liverwort *Pellia,* occur partly because of the tunnelling action of the water into the bank, and partly because the peat, unsupported on one side, moves slowly out over the stream.

Seepages or springs on blanket bog where water forces itself up to

The filigree petals of the bogbean flower may entangle visiting insects for time to ensure pollination. photograph: Richard Mills

the surface are generally slightly richer habitats. Iron-staining shows that this mineral at least is available. Rushes will often be found along with a sedge, *Carex nigra,* and perhaps the buttercup-like spearwort. Bogbean may grow in the water and the streaming stems of the floating clubrush give a new dimension to animal habitats. Perhaps most typical is the bog pondweed whose reddish green leaves lie on or below the surface. It is confined to very acid waters because it depends on carbon dioxide as such, which is found only in acid water or of course in the air. All plants need this gas to grow – they combine it with water to make their cell structure – but other plants are able to use it in the form of bicarbonate, which is the form it takes in alkaline or hard waters. The bed of a mountain stream is made up of large rocks, which make convenient stepping stones for crossing it. As time goes by, weathering disintegrates the bedrock, which can then be carved out by the scouring action of an ice sheet or by the stream itself. Journeying downstream, they jostle and scrape each other into finer and finer pieces so that only silt and mud may be left by the time the river reaches the lowlands. Any patches of sand or gravel that accumulate where the river slackens its pace do so only until the next flood comes. Nothing is static for long in a mountain stream: even the biggest boulder is being chipped away by the ceaseless onslaught of water loaded with sediment. Eventually, though it may take a hundred years, it will yield to pressure and begin to move downstream, creating a gap for another to fill. The sharpest rocks will be smoothed, as on a waterfall, the banks will be undercut and fall in and the bed will be remade by each flood so that animals and plants will continually have to find new homes and different sources of food. The upper Slaney in the Glen of Imaal, County Wicklow, is typical of this type of stream.

Picture a mountain river 2–3m wide. The water slides over sheets of sloping rock, it splashes over ledges and between boulders and rattles over rocks and stones. Around each object a complex community will be established, depending on the shelter provided against the current. Mosses and algae grow under water, other mosses and lichens above the surface. Where the rocks are constantly wetted by the spray of waterfalls, algae also make their slippery presence felt. Mosses are specially suited to shallow turbulent water because, like the bog pondweed, they can use only carbon dioxide, not bicarbonate. *Racomitrium aciculare* is a moss that will be seen in any hill stream, forming dark green straggling tufts a few centimetres long. Much larger, and familiar to many who visit rivers is *Fontinalis,* whose brownish leaves are placed in three rows on the stem. In the lowland it may grow almost a metre long but here in the mountains it is much shorter. A rather

e clear blue of the forget-me-not enlivens many a ıter course throughout the summer.
photograph: Richard Mills

different plant is the liverwort *Scapania*, which forms crowded spongy patches of bright green with its overlapping thin leaves. Liverworts are close relatives of mosses, but their shoots are often flattened and soft.

Moss-covered boulders provide a good habitat for clinging sorts of mayfly *Ecdyonurus* and stonefly *Isoperla* nymphs. (They are called *nymphs* because they look quite like the adult insect without its wings: a *larva* is a young insect that is totally unlike the adult form, as maggots are to blue-bottles.) Mayfly and stonefly nymphs are grazing animals which cat little of the moss itself but instead scrape off the algae from its surface and that of nearby stones. *Ecdyonurus* is flattened so it can cling to surfaces within the boundary layer of still water. It can escape the worst floods by squeezing into cracks between or beneath rocks.

Boulders create shelter, and on their downstream side more delicate animals can survive, such as the swimming forms of mayfly and stonefly. Also you may find a cased caddis fly larva, at first sight a little pile or tube of stones. The habit of building a case of stones or plant fragments is widely developed among caddises. The larva collects material soon after hatching and sticks it together into a tube with a type of silk from its mouth. Usually the case tapers, because as the animal grows larger it adds stones only at the head end. The hind end of the body has two large hooks holding onto the inside of the case and the insect walks dragging the case after it. As soon as danger threatens it can withdraw right into its case and pretend it is not there. But this gives it little protection against fish (particularly trout) which frequently eat the creature, case and all. More importantly, the case makes the caddis heavier and so less easily swept away. In faster currents a species generally makes a heavier case.

Holding station between the rocks in a mountain stream will be some small brown trout. They live a hard life in this habitat: food is scarce and they use a lot of energy just to hold their own against the current. They grow slowly and at five years old the trout of the Donegal or Connemara mountains may be only 20cm long. They are always ready to snap up terrestrial insects that may fall onto the surface and often they depend more on this source of food than on the aquatic animals which show themselves as little as possible. In slightly slower stretches the first dippers and grey wagtails may be seen, the dipper standing and bobbing on stones, or working upstream in the shallows looking for small animals. They hunt quite well under water bracing themselves against the current with half-open wings.

Life in Lake-fed Streams

Quite often in the mountains small valleys and hollows were deepened by moving ice so that water now accumulates to form a lake. On northern and eastern slopes, shading allowed snow to accumulate and ice to form. Moving ice plucked out rock to form a cup-shaped *coum* or corrie. Usually there is a high back wall and sloping sides and there may be a lip over which the ice had to pass in its slow movement to the lowlands. Coum lakes like Coumshingaun in Waterford or Lough Doon in Kerry are set in the bare rock. They are deep and dark and have few marginal plants.

River life is usually richer below a lake than above one, partly because some plankton washes out of the lake and forms food for animals. Also a lake evens out river flow as it can absorb some flood water (by rising) and release it over the next few days. For this reason algae are little damaged by floods and may grow to cover quite large areas. In summer their waving forms look grubby with the growth of other, smaller algae, but towards the ends the fresh green growth shows that they are green algae, more advanced than the blue-green forms. The green colour is chlorophyll contained in a chloroplast of distinctive shape. With a microscope you may see the spiral chloroplast of that textbook alga *Spirogyra*, the star-shaped one of *Zygnema* or the flat plate of *Mougeotia*. Even to the fingers or the eye there are differences: *Spirogyra* is very smooth and shiny, *Oedogonium* is woolly looking, *Bulbochaeta* rather tangled. These small algae may easily be dismissed as so much green scum, but there are different shapes – encrusting forms, short isolated tufts, felt-like masses or long trailing filaments.

Look again at the sheet of algae. There may be greyish stocking-like shapes moving with the ripples. These are the traps of *Hydropsyche*, the net-spinning caddis. Instead of sticking stones onto its silk tube this insect enlarges it greatly, attaching it to face the current like a wind-sock. The animal lies at the small end, and when enough debris has accumulated it ventures out and cleans it off.

Below a glacial lake a stream often meets sand and gravel left behind when the ice melted. The first higher plants may appear on these sediments. Water milfoil, its brown feathery leaves often coated by algae, is one of them, the bulbous rush another. This rush resembles many others when it is growing on *terra firma*: it has spiky green stems and small brownish flowers. When submerged, however, it grows long thin leaves that look like a grass and it never flowers, enough to confuse any terrestrial botanist. It is not at all the only case of such variation. Many water plants develop differently shaped leaves above and below the surface even on the one plant.

Waterfalls

Waterfalls occur in a river's course where there are sharp changes of slope. Ice action created many of these breaks in the hills: if it did not actually cut them it may have diverted a river to flow over hard ridges of rock that have not yet been worn down. Waterfalls change over time. In many cases there is already a gorge-like trench of rapids below a fall, showing that it is gradually working its way upstream.

Glacial valleys are U-shaped and often the ice so deepened the main valley that its tributary streams were left stranded at a high level on the valley side. These hanging valleys produce the waterfalls in Glendalough in Wicklow and Glencar in Sligo. Elsewhere, a coum eating backwards into a mountainside produced the same effect. Powerscourt waterfall in Wicklow falls into a small coum, while several small streams cascade down the walls above the Cloonee lakes in Kerry. The mountains of Kerry are in fact good places for waterfalls and rapids. Slopes are steep, water is abundant and the remains of a mountain glaciation frequent. The rivers gush through narrow gorges and turbulent pools perhaps 2–3m deep. The force of the water, which may rise another metre in a flood, would seem to make life impossible, but even here there will be a thin covering of moss, looking a bit worse for wear. Mosses attach themselves with tiny hairs much smaller and more efficient for this purpose than the roots of more familiar plants. Here is a good place to look for *Perla* the largest of our stoneflies. These may reach 4cm in length and take up to three years to develop from egg to adult. Having a hard skin like all insects it must moult at intervals to allow growth and *Perla* may moult thirty times. Unlike the great majority of stonefly nymphs it is a carnivore, so fishermen consider it a mixed blessing in a river. On the one hand it reduces the number of smaller organisms available to fish, but on the other it is itself eaten by the larger trout: it can even be used as bait.

Waterfalls above a certain height and bare rock faces form an impassable barrier to salmon and sea trout in their regular migration upriver to spawn. In a flooding lowland river they may jump 3m or more, but in the hills the turbulence and rockiness preclude such athletic feats. Below a waterfall there is generally a deep pool and, where the river escapes from it, some banks of gravel or sand. Here is a good place to look for the tell-tale depressions or *redds* left by salmon and trout after egg-laying. In fact redds may be found anywhere on the river where there is gravel.

Salmon and Trout

The salmon spawns in December, but the fish live in the rivers for much of the early part of the year. Some enter from February onwards in the 'spring run', others arrive at river mouths in late summer and work their way upwards when water levels permit. The oddest thing about salmon is that they do not feed in fresh water at all, though they can be tempted by the fisherman to snatch at a fly or lure, perhaps from force of habit. Some of the adult fish will not have eaten for ten months by the time they spawn. They are sustained by the rich deposits of fat in their muscles that make the meat so palatable to us.

By autumn, male and female salmon are easily distinguishable. Both have lost the characteristic silver flanks of the sea fish and have taken on darker colours to camouflage themselves against predators in the shallow waters where they will eventually spawn. The male's breeding colours are russet, brown, orange and red, all delicately blended into an overall coppery sheen. The male also develops a hooked lower jaw which may, like a stag's antlers, be used to threaten other males. It also makes feeding difficult if the fish were to attempt it. The females are almost black at this stage. Their bodies are swollen with developing eggs, which make up a quarter or more of their weight.

Salmon live in quiet places in the river through most of the summer. When there is high water they travel upstream and a flood any time from May to September brings fishermen hurrying to the banks. Here they see the fish moving up the rapids and weirs which are largely impassable at low water. By mid-November the breeding urge has become very strong and first the males and then the females begin the final phase of their migration and seek out suitable spawning beds. The homing instinct of the salmon is so finely tuned that the fish remember the taste of the water where they spent their first months and often return to almost the same bank of gravel where they themselves were spawned.

For some days before laying her eggs the female fish may be seen moving around gravelly areas of the stream. At times she flashes her tail, using it to dislodge bigger stones and generally test the ground. Instinctively she is looking for an area where water flow through the gravel is strong enough to oxygenate her eggs through the long four months of their development. Finally she is satisfied and over the course of hours or even days she excavates a depression in the sediment with her body, working downwards to 15 or 30cm. Spawning itself takes place around dusk. When the female is ready she rests her lowest fins against the bottom of the hole as if to test its depth. A male has by this time joined her, and after much quivering he stimulates her to extrude a stream of orange eggs into the redd. Simultaneously he releases a cloud of white sperm or milt which drifts downstream covering the eggs.

Often at this point one or two other males dart in and also release sperm into the redd. They are small fish, some of the young males that become sexually mature before their journey to the sea. They may be chased away by the larger fish, but they ensure maximum fertilisation.

All Pacific salmon die after spawning but some of the Atlantic species survive, more particularly the females, to return to breed a second or third time. Their long fast in fresh water is recorded by a scar on their regularly growing scales, and the rings on these scales also record such traumatic events as spawning and the change in habitat from sea to fresh water, so that biologists can piece together the life history of the fish by examining these scales.

The fish that survive spawning make their descent of the river in two groups, some immediately in December, others resting until February or March. They gradually develop a silver sheen and regain some of their former vigour. But these *kelts* are still characterised by large heads and a thinness about the body. Many will fall prey to predators and of the 40 per cent of adults that reach the ocean successfully, only one-eighth will return to breed a second time.

The redd is the home of the fertilised salmon eggs for the next three or four months. Water temperature is low at this time, so the eggs develop slowly, becoming 'eyes' within six to eight weeks. After the eyes, the backbones of the little fish take shape and during March the eggs begin to hatch and the *fry* appear. They are small, transparent, brown creatures at first, with protruding eyes. They still have a yolk sac suspended from their bodies for food. Gradually this is absorbed, so that by April and May the young salmon, or *alevins,* become active and leave their gravel home. After a few days they become territorial and space themselves out in the stream bed. The redds at this stage may have been modified by floods or filled in with sand or debris, but some are still recognisable.

The young salmon live in these headwater streams for their first two years. The alevins soon grow into *parr* with blotched sides and finally into silver *smolts* which migrate to sea in late April. Some return to breed the next year as *grilse* (perhaps of 2–3kg), others spend two or three years at sea and return at 8–20kg. It seems that our grilse feed in coastal waters off Scotland and the Faroe Islands. The others migrate further, to the areas around Greenland, where salmon are caught in large numbers. The proportion of these older fish in Irish rivers is currently decreasing, and grilse are becoming more common. Perhaps this is a response to fishing pressure on the high seas and to our own drift netting, but the loss of spawning grounds for the larger fish and their susceptibility to ulcerative dermal necrosis (UDN) are other factors. UDN is most severe in spring when the older fish come into rivers.

This split of the salmon population into single-winter-at-sea fish (grilse or peals) amd multi-sea-winter fish is one example of the peculiarity of many fishy life cycles. Intuitively we would expect all salmon or trout to behave in the same way, but here we have two strains of salmon wintering and breeding in different areas. Races of trout are even more separate. The original trout seems to have been the white or sea trout, which lives in the sea for most of its life but comes into fresh water to breed. Brown trout are those fish that fail to follow the call of their ancestry and remain all their lives in fresh water. They have lost contact with the sea trout and will eventually form a distinct species when they can no longer interbreed. The salmon has not gone quite so far, though grilse breed grilse to a large extent. A problem for those who manage fisheries is how to restore the older segment of the population, which is much sought after by anglers. Stocking with fry of this strain may be a partial answer in the short term, but there is some evidence that grilse cycles such as we are having at the moment are quite a natural and temporary feature of rivers.

The salmon fry and parr co-exist with other fish, especially the brown trout and white sea trout which also spawn in gravelly places. In flowing water they are all territorial, each snaps at or chases off intruders from its own patch. In this way trout usually dominate salmon of the same size. It seems that the object is to find shelter, not necessarily food. In still reaches of rivers and in lakes they get on well together and even form mixed shoals. Fish tire much more easily than mammals, so shelter from the current is essential to their well being. They face upstream for most of their lives to present their most streamlined profile to the current. Even the salmon smolts and kelts continue this habit in their journeys to the sea so they travel backwards much of the time, at least in the swifter sections of the river.

These young fish feed primarily on small invertebrates either from surface drift or from the water below. They are all opportunists, eating what is available and changing from one sort of insect to another as the seasons bring different sizes and shapes of prey. They also eat each other, though generally a fish has to be three or four times the length of its fellows before it can do this.

Drift

Surface drift is easy for us to see, but a lot of submerged drift also occurs, as you will see if you submerge a net in the stream. The number of animals moving in this way may be enormous, and counts of millions per day have been made in very large rivers, such as the Mississippi, particularly during floods. The animals that drift are mainly lighter species, like mayflies, fly larvae or freshwater shrimps. It is unusual to find many heavier snails or cased caddises drifting. More drift occurs at night, and a peak in activity usually occurs just after sunset. The crafty angler with a wet (submerged) fly may cash in on this dusk activity.

Only a proportion of the animals drifts in any night and they only move a matter of 50–60m downstream. All the features suggest that it is a natural part of behaviour and not simply due to chance dislodgement, though this may augment it. It has a function in spacing out the young of an organism that tends to lay its eggs all in one place. It also allows them to colonise new regions that have been denuded of life by flash floods or pollution.

Effects of Forest Planting

Conifer forests are now frequently planted around the headwaters of many rivers. Once there were natural woods of oak and pine but they were displaced by the spread of blanket bog and the activities of humans. Most of our stream life must have evolved in shady conditions, because before forest clearance the smaller rivers would have flowed through leafy tunnels for half the year. A deciduous wood allows a rich growth of aquatic plants and animals before the leaves open on the trees. Later in the year these leaves fall into the water in increasing amounts, dominating the habitat in October and November and providing a source of food for much of the year. A coniferous wood has a greater influence on stream life because of the constant shade. This keeps the water cool, but it also prevents plant growth all through the year. A few mosses are adapted to deep shade but they can produce little or no surplus to feed animals, and the fauna may depend on plant material brought in from outside, either from the banks or from regions upriver. As mentioned before, most rivers produce about half of the organic material required by their consumers, but this can fall to one per cent on a shaded stretch of channel. Bacteria and aquatic fungi occur here as elsewhere, growing on dead leaves and wood. There may be specialist feeders on them, which are favoured by intense shade, but the acid water draining from a coniferous stand inhibits their variety and their numbers.

Overleaf
The kingfisher is a bird more familiar in photographs than in the flesh, but it cocurs all over the country wherever there are small fish, clear water and bankside perches.
photograph: Richard Mills

Teal nest in heather along some of our rivers but they are much more frequently seen when they come to fee on older seeds washed in t the river's edge.
photograph: Richard Mills

The male swan is an attentive guard for his conspicuous mate on her nest on the canal at Naas, County Kildare. He will remain with her throughou the upbringing of the young
photograph: Liam Blake

When trees are planted in blanket bog the ground is ploughed to dry out the soil and give temporary control of heather until the young trees take over. Peat is exposed in the drainage channels and some flows into nearby streams. The escape of loose peat is perhaps the major effect of modern forestry in the mountains though loss of fertiliser may also occur, creating unusually rich conditions in unexpected places. The build-up of silt on gravel can smother the breeding sites of fish as has happened on the Inver sea trout fishery in Connemara. It is easy to forget that even the salmon may breed high in the hills where it can scarcely turn around in the stream channels. Another effect of mountain forestry is that stream flow fluctuates widely over the year because of drainage. Water is released faster than from a natural habitat, while during droughts the trees use up almost all the water.

Bird Life

Upland rivers have little bird life. You may see a good variety of birds when walking by the river but none, apart from the grey wagtail and dipper, will be dependent on it. The meadow pipits, wheatears and stonechats, the wren and the willow warbler will prefer drier sites on the bank or the valley side but a hatch of insects will bring them to the water's edge.

Sometimes on the slower reaches you may meet a common sandpiper, a long-legged wader stalking the gravel beside the bank or bobbing on rocks in the stream. Look out for them in hilly parts of the west, in Wicklow or in Antrim and, of course, on stony lakeshores. If they have a nest or young their anxiety will attract attention as they flit from stone to stone and even into bushes, calling all the while.

The dipper, by contrast, must be well known to most stream walkers. Throughout the year it lives on the same stretch of river and uses the same rocks in mid-stream to stand on. Often you can drive a dipper along the channel from stone to stone before it eventually gets tired and takes a long flight round behind you again.

Previous page
The contrasting colours of the dipper break up its outline against a tumbling stream and conceal it from its predators.
photograph: Richard Mills

Having waited patiently the heron has a speedy meal.
photograph: Richard Mills

Mute swans gather in flocks after nesting and often feed in estuaries and near towns.
photograph: Richard Mills

Sometimes it has reached the end of its territory, a matter of 1–3km of stream where it feeds and nests. It holds this territory even in winter and is a hardy bird whose numbers change little from year to year. People have seen dippers feeding under ice, and they begin building nests in February or March. The nest is a ball of moss and grass built in a cavity or on a ledge. It is often found under a bridge and sometimes in a natural site at a waterfall, even behind the curtain of water. The birds use the same site each year and often the same nest, though they may have to do repairs on it. It is worth looking under any bridge for a dipper's nest and listening at every bridge for its chuckling song. If there is not a nest in use there may be old remains or half-finished nests, for, like its close relative the wren, it seems to build more nests than it needs. If there are no dippers there may be house-martins' nests, the birds being attracted by the ever-present flies.

Algae in Shady Waters

Our river is now perhaps 5–10m wide with alternating shallows and deeper sections and small rocks spread out in the channel. Willows and alders grow on river banks (if the anglers do not cut them). They cast a dappled shade and create new conditions for aquatic life. Red algae like shade: both in the sea and in fresh water they tend to flourish best where light intensity is relatively low.

Lemanea is a large, rather seaweedy alga which grows a tuft of pointed filaments up to 30cm long in shallow water. The filaments have bands along their length, and early each year they spring up from a persistent base, almost disappearing when the water temperature rises. The spores this alga sheds into the water give rise to a creeping plant so unlike the parent that it was thought to be a separate species. Just when botanists had worked out the relationship, they realised – to add to the confusion – that another alga, *Rhodochorton,* was almost identical with the creeping form of *Lemanea.*

Red algae may look brownish or greyish green in their habitat, but they soon betray their identity in a plastic bag, especially if they are left there too long. They often have rather impressive names. *Hildenbrandia* is found on rocks generally in calcareous water. *Batrachospermum* resembles a necklace or a chain of frogspawn and it grows well in peat-stained water close to blanket bogs.

Life in Calmer Waters

Few rivers remain turbulent all along their upper reaches. Usually there are places where the gradient levels out and the water slips quietly between banks of peat or sand, carrying little rafts of bubbles from the rapids above. Changes of direction create bays and backwaters that are relatively quiet in the summer but take a greater flow in winter.

Here is the place to look for more variety in the plant life. There will be little algae, called diatoms, which are important throughout the river system, in the surface film of growth that covers all rocks and plants. In the quieter places they may spread in abundance onto silt and mud, forming a characteristic brownish felt, usually studded with bubbles of oxygen. Diatoms are only visible under a microscope, but their hard silica coats are sometimes beautifully intricate and worth making the effort to see. Some grow into star shapes or chains and they may break off and float freely in the water as plankton. They can give a slight colour to river water, and the Nore below Kilkenny is noticeably yellowish when it is joined by the bluer King's river. Other diatoms are cigar-shaped and these may move about slowly on a ribbon of slime like miniature snails.

Larger plants that grow also in lakes may appear in the river backwater. The water lobelia has a neat rosette of leaves at its base but when it flowers it sends a long ungainly stem up to the surface, raising its whitish flowers into the air before they open. The pipewort is similar in habit, but it has a much more restricted range in this country. It occurs also in Scotland and in North America. Most plants that grow on both sides of the Atlantic have been aided by humans in their travels; our indigenous pipewort is famous far beyond Ireland. Its roots are distinctive, since they have regular whitish bands across them. It is often worth looking through the drift of dead plants marking a high water line, if you are in Galway, Mayo or Donegal.

The fauna of this sort of channel is richer than the headwater streams. The sediments are varied and there is a measure of shelter. Large numbers of a small cased caddis fly often occur. These are the silver horn sedges, an angler's name for Leptocerids. The larvae build small cone-shaped cases of sand and these, unlike the adult flies, are eagerly taken by trout. The adults themselves have long antennae, hence the name. The mayfly *Caenis* may be similarly abundant and the fish frequently feed on them as they hatch in clouds in late May or June. It is not unusual to see trout hoovering up clumps of these insects from the water surface as they drop down to lay eggs. The fish are almost impossible to catch when this food is available and this mayfly has the descriptive name of 'the angler's curse'.

Water boatmen abound in the quieter reaches. They are fast-swimming blackish insects that row themselves through the water with the middle pair of their legs. They dart after their prey and also visit the surface to breathe. Take one out of the water into a jar, taking care it doesn't bite you, and you will see the glistening bubble of air caught under the wing covers. Leave it overnight in the jar and the chances are it will fly out in search of a more adequate habitat.

Breathing poses special problems for aquatic animals. Most of the insects that live in water are larvae. They have thin skins or develop gills to take in oxygen dissolved in the water. Adult insects under water are in trouble, for their thicker skins limit the amount of oxygen that can get through. If they live in turbulent water, as a few beetles do, they cannot visit the surface to obtain air. Instead they have a thin area of skin on their undersides which is covered with a myriad of waxy

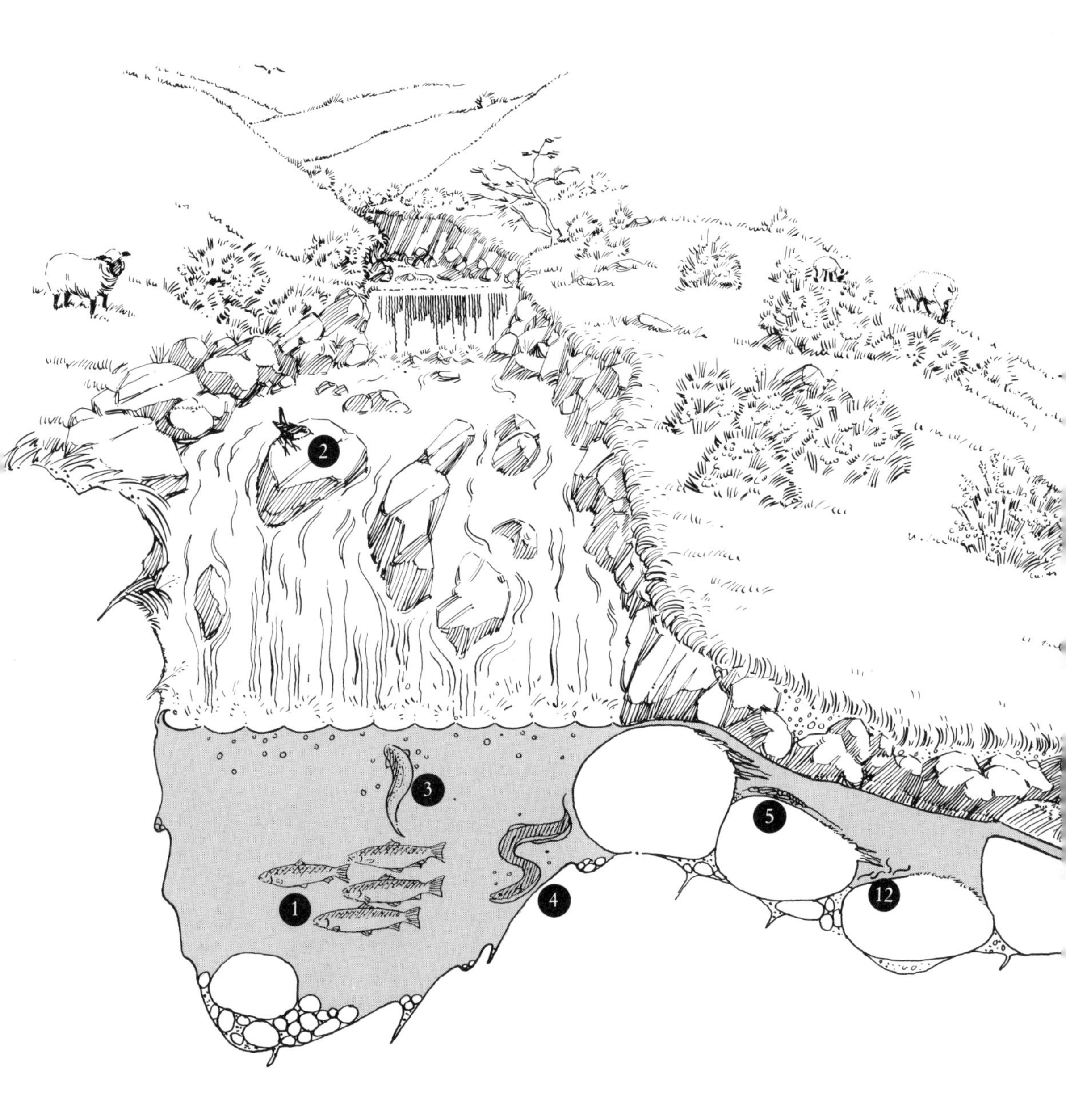

2
3
1
4
5
12

1. sea trout
2. dipper
3. salmon
4. eel
5. mayfly nymph
6. moss
7. fish leech
8. grey wagtail
9. cased caddis
10. free-living caddis
11. otter
12. midge larvae

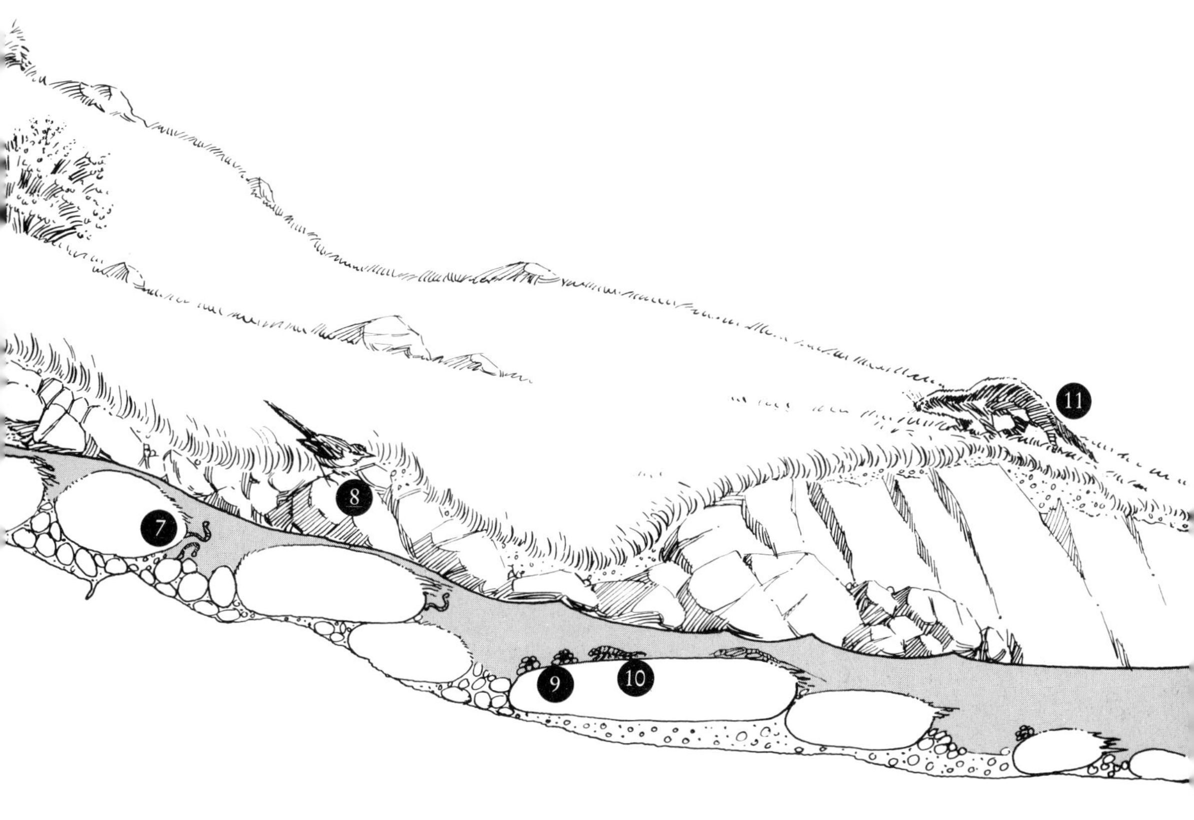

waterproof hairs. They can maintain a bubble of air here indefinitely and oxygen diffuses into the bubble from the water just as fast as the animal uses it up from the other side. The hairs are so small that there may be over a million per square millimetre of surface. In less rapidly moving waters adult insects like the water boatman may visit the water surface with impunity and this habit gives them access to as much oxygen as they need and allows them to be active and able predators. Diving beetles and the boatmen are some of the fastest swimmers among insects and can catch any prey that is small enough for them.

Water boatmen are opportunists, feeding on any animal that crosses their path. In this they resemble dragonfly or damsel fly nymphs, active animals that stalk their prey by feel or sight, shooting out a long pair of jaws when they are within range. At rest this structure is folded back under the head, looking like a sort of mask. Adult dragonflies and damsel flies have similar lacy wings, but at rest the dragonfly holds them out at right angles and the damsel fly folds them above its back. They both catch other flying insects and with their huge bulbous eyes they must miss little. They swoop after their prey like hawks, catching it in their legs and returning to a perch to eat it. Like birds they have some sort of hunting territory and chase other rivals away.

Calcareous Streams

Having followed this acid river from its source to the point where it is just about to leave the hills and flow into the lowlands we want to leave it for a moment and look at one of those rarer streams that happen to rise on limestone land, perhaps in the Burren region of Clare or the southern fringes of Fermanagh. They start life with certain advantages over acid streams because some of the nutrients required for plant growth have been dissolved out of the limestone and are already in their water, so they can support large plant and animal populations, even in their higher reaches. But there is another feature of limestone which is perhaps more important: it is soluble in rain water. Rain is naturally slightly acidic and, falling on limestone, it eats away cracks and hollows in the rock. This process is well seen in the Burren, where it has been calculated that the surface is disappearing at a rate of a millimetre every twenty years. This does not seem very fast, but projected through geological time it means that huge thicknesses of rock have probably been removed off the central plain and the western limestones. It is a wonder, in fact, that any hills remain unless they are capped by a more resistant rock.

Water sinks rapidly into the cracks in limestone and continues its work underground, excavating the pipes, caves and caverns in the rock that are explored by the potholer. Hill limestones are usually arranged in horizontal layers and water finds it difficult to penetrate each new bed. It may flow along such a surface and many of the seepages and springs occur here at the base of small cliffs or steps.

Life in Limestone Rivers

The flow from these rock crevices is intermittent and depends on recent rain. Quite often you will find only small pools, muddied by cattle and with no apparent flow. There may be marks on the cliff above showing that water flows there too when it has filled the porous reservoir of the rock. The Burren in a wet February looks very different to the dry desert of June. If the rocks in the vanishing streams are shaded, mosses are likely to be abundant. The ones towards the centre of the channel will have a ragged look, their lower leaves damaged by the water. Aquatic animals are rare because of the danger of drying up. Mites and springtails invade the moss from the surroundings, woodlice wander into it and flies will rest there. But truly aquatic species are almost absent. Perhaps on the surface of the pools there will be some large, long-legged insects, the powerful-looking pond skaters or the water measurer, almost unbelievably thin. Both are bugs, a group of insects with long, needle-like mouthparts ready to suck the juices out of some hapless prey that falls onto the water. They make use of the surface film on the water and flick themselves about with their legs.

In the pools themselves tiny thread-like creatures only 2–4mm long are often seen thrashing around in an S-shaped curve. They are the larvae of the biting midges. During the late spring and summer several generations of these flies may appear, giving ample opportunity for some of them to obtain the meal of blood that they need to lay eggs. Their growth is rapid and the life cycle from egg to adult may take only three weeks, though in the autumn the eggs remain dormant until the following spring. These animals are well adapted to shallow pools, as they can withstand high temperatures in the water.

Temporary streams on limestone usually become permanent as they descend to lower levels though it is difficult to generalise about water flowing on the surface of such a spongy rock. Sometimes whole rivers sink into the limestone, even reaching the sea without ever reappearing. The Caher river in the Burren is the exception, since it flows from source to sea above the ground and plainly visible. The clarity of its water is noticeable: there is no peat-staining and the water has a metallic brightness about it. But feel its temperature too, for in the summer spring-fed rivers are often cold to the touch, while on some winter days they may be warmer than their surroundings.

Steep slopes are seldom part of limestone country for any great distance and the stream is likely to have gravelly patches near the bank. Rather unexpectedly, bluebells may grow here; they are able to persist here because of their deep bulbous root and the generally low water levels in the spring. Their presence shows that these banks are seldom eroded. Spates are reduced in limestone country because there is such good storage under the ground. The bluebells also tell of a certain richness in the environment, as do the numerous land plants nearby, quite different to the situation on moorland.

Spring water is sometimes slightly acidic when it rises and so the upper course of a limestone stream may be deeply cut into the rock. The acid is quickly neutralised by the lime, however.

The stream valley is often now filled by hazel scrub, alive with the songs of blackbird and cuckoo in early summer. Head-high scrub is difficult to penetrate at the best of times and very easy to get lost in. However, cattle often make tracks through it, drawn by the sounds of running water. The scrub itself is a damp, cool habitat in contrast to the rocks outside. The branches are heavily clad with mosses and lichens and down at stream level there may be other springs and muddy places. The stream itself is moderately fast and 3–4m wide. Its bed is made up of gravel, rocks and boulders. The larger submerged rocks are covered by *Fontinalis,* the large moss which is as much at home in this alkaline water as it is in acid. The long strings of the moss have lost most of their lower leaves to the current but in their place are small gelatinous beads. These are *Nostoc,* a blue-green alga embedded in a tough jelly. It is a very widespread species: it grows also in the little water-worn hollows on the limestone surface, on old asbestos roofs and on shady patches of ground, even on lawns. It is crusty when dry and rubbery after rain. Often on land, *Nostoc* is lichenised. It has been discovered by a microscopic fungus which has become partially parasitic on it. The alga supplies food by photosynthesis, the fungus scavenges for water and mineral salts and they both steal each other's surpluses. The two plants lose their separate identities and together form a lichen.

Freshwater snails do not e[at] water plants but rather use their strong tongue which has a sandpaper type surfa[ce] to scrape off epiphytes and algae from rocks and plan[ts]
photograph: Richard Mills

Insect Life in Limestone Streams

The animals of small limestone streams, such as the Little Brosna or some of the Suir tributaries, are similar in type to those in acid waters, though richer in numbers and variety. There are stonefly nymphs with two tails and mayflies with three. The stoneflies of a hardwater stream include *Nemoura* and *Leuctra.* The adult female which, unlike the mayfly, folds her wings flat on her back, scatters her eggs on the

A rather fiercesome little fish the male stickleback w[ill] courageously defend its ne[st] against all intruders during the spawning season in spring. The strong sharp spines along its back and sides do not seem to deter natural predators such as trout who frequently feed o[n] these small fish.
photograph: Richard Mills

Anglers' Rest
Tavern

surface of the water and they soon sink to hatch into slow-moving nymphs. These tend to creep over the gravel and rocks rather than run or swim like mayflies, and are known to anglers as 'crawlers'. They usually live only a single year in the water and, come the following spring, they climb up a solid support into the air, the nymphal skin splits and the adult insect flies away when its wings have hardened. Most stoneflies like clean upland rivers and, since they usually require water rich in oxygen, their absence may suggest pollution.

Mayflies resemble stoneflies to some extent, both in appearance and life history. When fishermen speak of mayflies they mean just the species *Ephemera danica*. To others of the group they give colourful names like ambers, spinners or duns. The nymphal stage of these insects may last from three months to three years depending on species. Once they have hatched into adults they can no longer feed so they are truly ephemeral, as their Latin name suggests.

\ low river can often force e trout angler to resort to a pipe of tobacco and the ımble worm. Waiting for a bite on the Little Brosna near Birr, County Offaly. photograph: Richard Mills

The change into the adult fly goes through two separate stages: they emerge from the water as a *subimago*, a colourless copy of their final form, and they moult again almost immediately. If you see one adult mayfly you will see a hundred. This synchronised emergence means that they so outnumber their predators that some always survive to perpetuate the species.

Mayflies are of exceptional importance to the trout angler because their habit is to cling to the water surface before emerging as flies and also to fall back into it after egg laying. Fish take them greedily and may take the lure of the fisherman just as easily at this time if it is realistic enough. Fly-tying is both an art and a science. As well as copying the shape, the angler also mimics the behaviour of the insects, dipping the fly onto the water surface or resting it upon it. Real mayflies float downstream with their wings held vertically and so look like miniature sailboats. The fisherman's do the same.

The habitats of mayfly nymphs are very varied. Some cling to rocks or moss, while others swim and a few burrow. In the calcareous stream there are often some flattened nymphs of *Ecdyonurus disper*, August duns. Like all mayflies they have a series of flaps down each side of the body which pulsate whenever the animal is active or frightened, as it will be if you take it out of its habitat. The flaps are called gills and while they take some oxygen from the water they also bring a current of fresh water into contact with the insect's underside. The speed of the beating varies with oxygen content and with current speed so it is a sort of panting used only when necessary.

sting in a typical pose this salmon parr appears to be sitting on its pectoral or front fins. The distinctive thumb markings along its lanks and the rather more silvery appearance ifferentiate it from a trout of similar size. otograph: Éamon de Buitléar

At the very base of the moss stems and amongst the algae stuck tightly to the rocks are a myriad of small worm-like creatures. These are larvae of the non-biting midges – a cumbersome English title for the Chironomids. They form a huge and very numerous family and there is hardly a freshwater habitat that does not contain its quota. Of all the fly larvae they are among the best adapted for an aquatic life. Different species live amongst vegetation, in sediment and beneath stones. They are important consumers of detritus and in turn form a major food source for all carnivorous animals in fresh water. In adult form they are delicate greyish or greenish flies, which are attracted through an open window to light but usually perish in the heat of the bulb. The males often have noticeably feathery antennae.

Lime Deposits

Just east of Rahasane turlough (a turlough is a dry or disappearing lake) in Galway the Kilcolgan river is a clear rippling stream but it soon enters a gorge. Steep cliffs in limestone country have normally been formed by the collapse of rock into the underground cave and ice has often carried the loose rock away. Such a river gorge is now floored by solid sheets of rock and the water spills rapidly over them.

Seepages at the edge of the gorge flow into cushions of moss and the new water drips off their leaves. Touch one of these cushions and it will be surprisingly hard, coated by a layer of lime from the water. The lime is dissolved underground by water – such as rain water – that contains carbon dioxide. When this gas is removed by plants or by turbulence the lime reappears in solid form. So mosses, liverworts and algae by their very growth bring about a deposit of lime that is far from good for them. They must continue to grow outwards while their lower, older sections become set in a sort of concrete. The deposit itself is called *tufa* and there are sometimes small amounts on bridges and walls where water has seeped through the limy mortar.

Lime deposition occurs around springs but also on the river bed or waterfalls, where turbulence causes a loss of carbon dioxide, or where algae extract it from the water. If the stones are small enough they may be rolled around the stream bed and form little balls, spongy in appearance but hard to the touch. The algae which create them are often blue-green types such as *Rivularia*. They grow in many parts of limestone streams and form gelatinous but tough colonies like beads. Tufa forms especially in the low water conditions of summer when temperatures are high. As the stones dry out they come to look white and powdery as do algae left stranded on a water plant.

Limestone rivers more often run from springs than do acid rivers, because the rock is porous. Sometimes the springs are huge, giving rise to a river all at once. The Cregg river in Galway, a tributary of the

Clare-Galway, begins its surface life like this, forming a channel up to 15m wide. It is fed by an underground stream and it bursts out of fissured limestone in an open field. Its bed is encrusted with lime deposits as would be expected, and its fauna is sparse. Watercress lines the channel and invades any minor springs at its edges. It is a straggling plant with white flowers and leaves that are often bronzed in winter.

Plant Life in Limestone Streams

On clean rocks or slabs in the rapid flow you may see distinctive patches like dark red paint. These are colonies of the red alga *Hildenbrandia,* growing so closely on the rock that they remain in the boundary layer and out of the current. *Hildenbrandia* grows slowly, in fact too slowly to survive on softer calcareous rocks like chalk. It has a slimy feel and this gives it an advantage as it prevents other plants becoming attached to it. Sometimes you may find the larger green algae so coated with a skin of small relatives that they can no longer grow. The lichen *Verrucaria* has a similar life form to *Hildenbrandia* and it forms faint green patches on submerged rock.

Submerged moss falls off in abundance more rapidly in a hard-water river than in an acidic one because of dependence on dissolved carbon dioxide. However, it may appear again lower down the river if weirs or other obstructions cause rapids or where the decay of organic matter releases this gas. The mosses still grow on the banks, however, and on an undercut rocky bank liverworts too will generally be found. One of the most distinctive is *Conocephalum* which grows broad green ribbons marked with pale spots. Break a portion of the ribbon and you will get a pleasant smell, reminiscent of soap. Liverworts, like mosses, reproduce by spores and the sporing structure in *Conocephalum* is like a little parasol about 6cm high. The spores are shed from capsules beneath the canopy and are carried by the wind to new sites.

Below a series of rapids or a waterfall there is generally a stiller section where there is some accumulation of sand and gravel. A number of organisms are just waiting for this kind of habitat to establish themselves in and they include some flowering plants. We tend to think of these as 'real' plants, but algae, mosses, liverworts and ferns are every bit as much plants and they far outnumber the plants that flower. We have already travelled a good distance downstream without seeing any large plant rooted in the water. The banks have been clothed in vegetation of several sorts, woodland, grassland and marsh, but the stream channel itself has been bare of anything larger than mosses. This is not because the water is hard or lacking in nutrients, but is due simply to the current.

It is perhaps surprising that so few flowering plants have become adapted to life in turbulent streams: in fact there are only two groups of tropical plants that favour this habitat. Plants have managed to colonise practically every other site from the driest, wind-whipped desert sands and poorest tundra to tree bark, rocks and shallow seas. It may be the variation in conditions that they cannot take, being bombarded by stones and floods at one moment, being left high and dry by a low flow in the river the next. Also, and perhaps more importantly, most plants cannot set seed under water. Many fail to flower at all if the water is too deep and most have to send their flowers into the air. This must be a real disadvantage to the species and inhibit its ability to evolve over time.

Aquatic plants first appear fringing the channel, growing in the no-man's-land that is flooded in winter but dry in summer. In a calcareous stream the species may be fool's watercress, *Apium nodiflorum,* and the small water parsnip, *Berula.* Both these are in the parsley family and so have characteristic tiny flowers set on the spokes of an 'umbrella'. The bright blue brooklime may occur too and there will usually be sweet grass, *Glyceria,* covering muddy places on the banks and extending out onto the water with its wide, floating leaves. The presence of even a few plants slows the current and attracts sediment, thereby creating a whole new set of habitats for animals. To a small animal the value of a water plant is as a habitat not as food. You will seldom find a water plant (unless it is a yellow water-lily) that has been eaten, in direct contrast to terrestrial plants, which are beset by snails, slugs, caterpillars, beetles and grazing mammals.

Water crowfoots seem to us one of the best adapted plants for stream life. They are really white buttercups but their leaves are different from their land relatives. The submerged ones are tassels of long hair-like parts that offer little resistance to the flow of water. In a fast current the leaf folds up, its segments pressed in on each other, but when flow decreases they open out and are held like brushes. When they flower they tend to produce a few floating leaves, possibly to keep the flower up, and these are more common in species that grow in still water. Indeed the few species that grow on the mud around springs and in ditches may have only this sort of leaf. The water crowfoots characteristic of swift streams form mounds of vegetation rapidly every spring. Roots sprout from anywhere on the stem, the clumps collect silt and may narrow the channel. In this way they speed up water flow and may themselves be torn away before they flower.

4 Lowland rivers

Hill streams run in V-shaped valleys constricted by the terrain in which they find themselves. They plunge into ice-cut hollows, they squeeze past outcrops and rocky spurs and they tumble from boulders and ledges. When they reach the lowlands they escape these confines and are freer to exert their own influence on the landscape. The deep places and shallows, pools and rapids that alternate with each other in any stream are only partially caused by rocks and obstructions in the river bed – they are also a feature of water flow itself. Run the water from a hose down a slope of soil and you will see such shoals develop naturally. As the speed of water increases so the friction on the river bed also rises until it disrupts the smooth flow of water and causes turbulence. Sediment is deposited here in the slower water. So it happens that alternating shoals and chutes develop at intervals depending on the speed of the river, which in turn depends on the gradient of the channel.

In lowlands there are fewer obstacles to prevent this characteristic from showing itself and the river begins flowing in those great looping bends called *meanders*. Water speed varies across the river. It is slowest at the inside of a bend, so this is where deposition occurs and there is often a beach of stones with pieces of driftwood lying about. The river is shallow, but it deepens rapidly at the other side where erosion is being caused by the faster currents on the outside of the bend. The shoal often extends out into the river in an arc, which deflects the flow across to the eroding bank rather than straight down the channel. It also creates a *riffle* – a relatively shallow, more turbulent patch of water. There is an eddy current in many cases on the deep bank and you can watch the floating bubbles or other debris circling slowly in this whirlpool.

Once a river has started to meander it continues to do so and it is a major engineering exercise to try and stop it. The bends themselves tend to migrate downstream undercutting the lower bank of each bend. It is a slow process, measured in centuries, but the meanders cover all of the broad valley floor, working through the sediment that the river itself has deposited in the floods of former years. The meanders do their bit to widen the valley when they touch its rocky edge.

Moving from the rocky reaches of an upland river to the slow bends of the lowlands will show major changes in plant and animal life but there is seldom an abrupt transition. The physical features of a lowland river, the meanders and the slow flows, may appear anywhere, even on a mountain plateau, but the nutritional change is more usual at low altitudes. Most of our lowland rivers run through a glacial drift soil that is rich in nutrients. Thus any deficiencies that have been apparent in hill streams are soon made up. There are abundant supplies of nitrate and phosphate, calcium and potash, and there is also input from human activities in towns and farmland.

A few rivers, however, retain the acid, nutrient-poor water they received in the uplands right down to their mouths. Mostly they run on the seaward side of the mountain and are relatively short and fast. Such acid rivers flow in Connemara and Mayo, for example the Erriff and the Owenduff, and in Kerry, the Caragh.

Plants in Acid Lowland Rivers

Close to the hills or to other rock outcrops their channels will be very mixed with every size of material from boulders to fine sand. Water crowfoot may be found on the riffles and in shallower places, growing in fairly small patches.

You may find water dropwort too, especially on a stony river or on the bank above. It is a fine upstanding plant, bearing large leaves divided into many small parts, like an uncurled parsley. Like parsley it is an umbellifer and has been mistaken for wild celery, with fatal results. Celery, however, only grows in salt marshes. The water dropwort, *Oenantha crocata,* lines the banks and edges of most rivers off the limestone in Munster, Wexford, Down and Tyrone.

Submerged in the water in quieter places there may be some water starwort, its unpretentious but pretty rosettes of leaves growing towards the surface and sometimes floating on it. Like most water plants it absorbs nutrients through these leaves as well as through the roots. The leaves are thin and have no shiny cuticle to prevent such exchange, which may in fact be more important than root absorption. The leaves are constantly bathed in the fresh supplies of water that flow past, whereas the roots have to grow into new sediment to find more nutrients. Unlike most water plants, the flowers of the starwort open under water though 'flowers' is a rather flattering name for the single stamen or style that the plant produces. However, the pollen gets from male to female flower, the system does work and plants are found with seeds, sometimes in large numbers. Indeed, despite their humble appearance, water starworts are quite highly evolved plants, as underwater pollination is most unusual.

Insect Life

Animal life in these acid, nutrient-poor rivers such as the upper Liffey, the upper Dargle in County Wicklow and the river Bush in County Antrim is very varied, but not usually very dense. A variety of cased caddis larvae is usually found, the most distinctive being *Agapetus*. This larva makes an oval case out of coarse sand, which looks like a tiny igloo. It may be stuck down onto the rock, but this is only when the animal is pupating. When it is feeding it drags its case unhurriedly from place to place looking for new feeding grounds rich in algae. A similar life is led by *Sericostoma personatum*, the 'Welshman's button'.

In these stiller waters you will often find swimming mayfly nymphs. It is difficult to see them swimming in the river itself, but if you put one in a glass jar you will see how it flicks its body up and down with its three tails spread out. In some species the tails are fringed with hairs, which must add to their propulsive effect. The olives *Baetis* are common swimming mayflies, very abundant on a stony river bed with moderate flow. Their bodies are cylindrical rather than flattened and are in fact the perfect streamlined shape, broadest about one-third of the way back and tapered at both ends.

Looping slowly along the bottom of such a river you may also see worms like earthworms or the longish grubs of craneflies (daddy-long-legs). Their terrestrial relatives are the leather-jackets, which can wreak havoc on the roots of grass or cereals. In water the larvae usually remain buried in sand or mud, moving through it and eating detritus as they go. They hatch into adults in late summer and, being weak of flight, they are often swept onto the water surface and into the mouths of trout.

Eels

The eel is a sea fish which spends most of the middle part of its life in fresh water of all kinds. It spawns in the Sargasso Sea, 5000km from the parts of Europe and North Africa where it grows up. The larvae that hatch from the eggs are transparent and leaf-like, flattened from side to side. They live in the upper layers of a tropical sea that is 6km deep and they feed on tiny floating algae. The Sargasso Sea is a sort of giant whirlpool off the West Indies, a collecting ground for huge quantities of seaweed and a place dreaded by the sailing ships of old. From its northern edge the prevailing north-easterly winds move some of the surface water off into the Gulf Stream current which flows across the Atlantic to our shores. On this current the little eels are carried, but many must fall prey to fish and birds on the three-year journey.

The ones that survive change into *elvers* in coastal waters. The elvers, which look like miniature eels, enter Irish rivers in March or April. Many of them move towards the headwaters. They are athletic little fish 5–6cm long and they can wriggle up waterfalls and rapids and also move, snake-like, over wet grass in ditches. Their rate of growth from then on is slow and depends on both food and temperature. By their first Irish winter they average 8cm in length and add a further 10cm the next year.

The eel is a bottom-dwelling fish, but it is far from the slimy, mud-eating creature of legend. It is a strict carnivore, eating fish, insect larvae, shrimps, snails and frogs, mostly by night. Its shape allows it to glide through water weeds and among stones. It moves in a continuous wave and then stops, sinking onto the bottom. Most fish are buoyant: they flick their tail or their fins a few times and then hang in the water for a moment without moving; but bottom-dwellers like the eel have lost this buoyancy, which would greatly interfere with their lives, and it is surprising how quickly they sink.

Eels feeding in fresh water are called yellow eels but eventually they start to mature into the adult form, the silver eel. They are now anything from seven to fourteen years old. (In exceptional cases they reach fifty years if a barrier prevents them returning to the sea.) The skin becomes dark on the back, silvery on the belly, the eyes increase in size and the jaw muscles and intestines shrink. Gradually the eels stop feeding, and some dark autumn night when the river is in flood they travel seawards in vast numbers.

Large eel fisheries are run on the Erne at Ballyshannon and the Shannon at Killaloe, while the eels on Lough Neagh are caught in the lake or at Toombridge as they enter the Lower Bann. Once at sea, the eels disappear. A few have been caught at sea, but presumably if they do not need to feed they can travel at some depth. It is thought that they swim westwards, back to the Sargasso Sea, perhaps taking the southerly route near the Azores where the currents are more suitable.

There are other small fish in the river that are superficially eel-like. These are brook lampreys, relatives of the sea lampreys which attach themselves with suckers to large fish and feed off their blood. Our river lamprey is not such a parasite. It uses its sucker only to anchor itself to rocks and to feed on invertebrates of all sorts. It grows to about 15cm.

River Birds

One of the commonest birds of the river is the grey wagtail. In our journey from the hills we have heard the metallic 'chink' of its flight

Overleaf
Damsel fly and dragonfly adults adopt vivid and distinctive colours. Notice the dark patch on the fron[t] edge of the wings. It is thought that this pterostigmata helps the insect judge the widths of openings when flying.
photograph: Richard Mills

A river flowing through mountainous limestone ca[n] often have a dishevelled appearance. Here at the river Fergus in County Cla[re] the softer rock is dissolved away to form craggy pool[s] and gullies. The clear wate[r] rich in nutrients harbours [a] luxurious growth of plant[s]
photograph: Liam Blake

call many times and perhaps seen its deeply undulating flight. On the ground, as it flits from stone to stone or runs in under an overhanging bank to catch an insect, its long tail is clearly visible, especially as the bird moves it up and down all the time. Grey wagtail is something of a misnomer: the underparts are as yellow as anything, and were it not for the existence in Britain of a species called the yellow wagtail, our bird would doubtless have this name. Grey wagtails have a light bill. They can only deal with small food organisms and they must be alert at all times for passing insects. They catch these alive or dead, along the water's edge. Even when they have no young to provide for they spend much of the day feeding, and this activity draws attention to them. Contrast a kingfisher or a sparrowhawk, which after one or two good hunting trips can spend the rest of the time motionless and concealed.

Grey wagtails suffer in cold weather if they can no longer pick up food beside the water. They often come into towns in the coldest weather and some migrate to France and Spain. In early summer they build a nest in a cavity behind ivy or on a wall or bridge. It is generally near flowing water but also can be several hundred metres away. The liquid, chirruping song of the male can then be heard to advantage.

Herons also are ubiquitous. Anywhere there are fish there will be herons, though more occur on lowland rivers where the feeding is easier. Herons seem perfectly adapted to their habitat. Their long legs allow them to stand even in quite deep water. Their head and neck are light, and with their main body weight far back they can move them very fast to strike equally fast-moving prey. Herons have been known to spear fish on their beaks, but this is probably by mistake: they usually grab them from behind. Their flight is leisurely, on broad rounded wings, and they fold their necks back into their shoulders for better balance.

Previous page
The elongated body of the damsel fly makes mating difficult but not impossible as this photograph shows. Its extraordinary looped mating behaviour can be observed on bankside vegetation throughout the summer months.
photograph: Richard Mills

As they fly they are sometimes mobbed by crows, thrushes or tits, which treat them as they do birds of prey. It is difficult to see why this should be so, as herons never catch such small birds. They may take young moorhens occasionally, but generally their diet is fish and frogs. The heron itself seems a bit bemused by such attention from other birds – it cannot know that its wings have some resemblance to an owl's.

Adult caddis fly can often be confused with household moths but their roof-shaped wings and fine hair rather than scale make their identification relatively easy.
photograph: Richard Mills

Only when herons are nesting do they seem badly designed, as they perch ungainly in the tops of swaying trees or attempt to build their huge nests of sticks. Even cormorants appear better adapted to an arboreal life, which is not saying very much. This large black diving bird comes far up river to fish in winter and spring. It flies high over the valley like an immense goose and it lands in trees, especially dead ones, to roost and spread its wings. It also used to nest widely inland on the larger lakes but it is now restricted to a very few, mostly in the western half of the country.

Fish Farms

Herons and cormorants are quick to take advantage of fish hatcheries and farms and are always very unwelcome visitors. Fish farms are often located beside clean soft-water rivers because the algal growth typical of richer waters would be a nuisance. The farm diverts part of the flow into its various ponds, which hold fish of different ages. The water brings in oxygen for the fish and it also carries away their wastes which would otherwise build up rapidly because of the large numbers present. The fish are fed on high-protein pellets and grow fast, to be 350–500g at one year old.

Rainbow trout were originally introduced into Ireland from the United States at the turn of the century. It was intended to use them for stocking sport fisheries, but this largely failed as they do not adapt well to our rivers. They can be used as 'put-and-take' fish, however – released at full size to be caught again immediately. This is being done on a small scale with our present stock, which was reintroduced in 1955.

The rainbow resembles the brown trout to some extent but it is spotted on the tail and gill covers. It also has a pinkish band along its sides. Fish escape from fish farms sometimes, often when floodwaters overfill their ponds, and they can survive for some time in rivers travelling a few kilometres. They normally die within two or three months, as they fail to adapt to natural feeding. They become used to being fed on pellets at the surface and cannot recognise the natural food around them. Instead they take feathers, stones, plant stems and paper – all of which have been found in the stomachs of escapees. Sometimes, in richer rivers, they may be able to survive for several years, but they fail to breed successfully. Recently it has been shown that both brown and rainbow trout loose their ability to spawn properly when they have been bred artificially for several generations.

Freshwater fish farms can be commercial enterprises, though the margins on bought-in food are narrow. Rearing rainbow trout in sea cages produces a better-tasting fish in a shorter time, and since the methods have now been worked out this type of farming may come to supplant the other. This may be no bad thing, for fish farms can cause river pollution in dry weather if they are badly sited or designed. To keep the fish alive the farmer may have to take a large part of the flow of the river. In the sea greater volumes of water are available to carry away and break down the wastes.

It often seems that soft-water (acid) rivers are the preferred home of salmon and brown trout. Such rivers often run in hilly regions, where turbulence fully saturates the water with oxygen and produces good supplies of gravel for breeding. But this preference may be more apparent than real. It may be that coarse fish like pike and perch displace salmon and trout from the calcareous rivers they would choose if left to themselves. Certainly the largest brown trout come from hard waters such as Lough Sheelin, the Fergus and the Inny, and the largest salmon used to come from the Shannon. Sea trout, however, do prefer acid rivers (the Gowla in Connemara and the Dargle in County Wicklow, for instance) though they are not confined to them.

Sea trout (white trout) are plentiful in most short rivers running directly into the sea or into a coastal lake. They have a similar life history to salmon, but since they are smaller they can negotiate shallower water courses. The adults lay eggs in riffles in the head-waters, and the young fish usually spend two or three years in fresh water before going to sea. When they leave the river mouth they do not migrate far and may feed in shoal estuaries and along beaches, taking sand eels, herring fry and crustaceans. At this time they may be found in shoals offshore, weighing up to about 200g. Like salmon, some of the fish return to the river within a year, running upstream as *finnock* in the summer months; others spend a few years at sea before spawning. So there are usually some larger maiden fish in the run. Sea trout return for several years to breed in the same river.

Calcareous Rivers in the Lowlands

Having followed the soft-water river down to a point where human influence begins to be important, we shall now visit the middle stretch of a calcareous river like the Shiven in County Galway or the Suir in County Tipperary. There are few such rivers anywhere in Europe that are still in anything like their natural form. Flowing through rich farmland their banks have been populated for a long time and they have been subjected to all the impacts of agriculture and drainage. But we do have a few examples left: there are sections of the Clare-Galway and even the disappearing Fergus that would qualify.

Plant Life in Limestone Rivers

Green algae, especially *Cladophora* will be a natural feature of such rivers with a moderate flow. *Cladophora* occurs in two forms, one thread-like with long tough filaments, the other brush-like with short, branching stems. The brush forms collect fine silt and eventually become too large and so are broken off and swept downstream. When *Cladophora* spores germinate they grow into a generation of plants identical in outward form to their parents but with half the number of genes in their cells. Spores from this generation must fuse together before they grow into new plants like their grandparents. This alternation of generations occurs throughout the plant kingdom. It is more common in algae for one of the forms to be much smaller and of different shape to the other, and this is the situation also in mosses and ferns. In higher plants this stage is so small that it stays attached to the parent inside the flower or cone. Its spores come together when the pollen tubes meet the ovary and they fuse to form the seed.

Along with *Cladophora* on the riffles are usually some mosses taking advantage of the carbon dioxide trapped in the moving water. You will find the familiar *Fontinalis,* impartial as to acid or alkaline water, but there may also be long stringy clumps of *Eurynchium riparioides* with smaller leaves. This plant grows low down on stones or on the banks so that it is temporarily or permanently submerged.

Gravelly areas may see the first clubrushes growing in the channel. The dark green stems are used in rushwork: they fade to a light brown after a year or so. Here in the river they grow in an underwater form in mid-channel and produce stems only towards the edge. The submerged plant has long strap-like leaves that grow in tufts and sway in the flowing water. In shallower water it dispenses with leaves altogether and sends up its round stem clear of the water. These bear a few small brown flower-heads at the tip. Grass-like leaves must be efficient in flowing water, for several other plants also adopt them. A bur reed *Sparganium emersum* is one which, cunningly, often grows amongst the clubrush and is apt to be passed over. Its leaves, however, are more spongy and usually lie partly on the surface.

Exploring Deeps and Pools

These plants grow in water about one metre deep in summer. They are difficult to examine and hard to collect on foot. You could go after them in a boat, but leaning over the side of a boat, especially a narrow one like a canoe, is one sure way to capsize it. You can throw grabs in off the bank, hauling them in to see what has been snared. But the best way of investigating the deeper river is unquestionably by snorkelling. A facemask opens up new worlds under water: it makes the blurred colours spring to life, as good as any photograph. Fish treat the snorkeller with equanimity, shoals move to the side to let you pass and then return to their stations. Plants appear in three dimensions, showing their role as habitats for snails and caddis larvae crawling over their stems. Even the kingfisher may come to investigate you for a moment, though it will retreat in a flash of colour from the walker on the bank. The happy snorkeller wears a wet suit, but in July and August water temperatures are high enough to do without one, except perhaps in the case of a spring-fed river.

Pools of 3–4m deep with sandy floors may be empty of large plants, especially if they are shaded by trees, but where light can reach there will be the pondweed *Potamogeton*. Their large but thin leaves may be arranged untidily, as in the shining pondwood *P. lucens*, or in definite rows. Those of *P. crispus* are curled daintily along their edges. The large submerged leaves of the yellow water-lily may be a surprise. Everyone knows the circular floating lily pads, the homes of countless frogs in fiction, though not in fact, but the plants also produce light green, almost transparent leaves like rhubarb, which never come to the surface. Sometimes you will see the effect of the current on these as the larger leaves grow on the sheltered downstream side of the plant.

If you have not met with stoneworts higher up the river there must be some here as they are very characteristic of calcareous rivers. They are large green algae which look so like higher plants that they are often appropriated by the writers of floras. They have stems with rings of branches, a bit like a horsetail, and they are greyish in colour. Often they grow in large lawn-like groups which can be very beautiful. The grey colour is caused by lime, which is deposited on the outside of the plant and may form 60 per cent of its weight. The smell of the plant is no less distinctive than its feel, and if you take some out of the water it may remind you of old fish.

The light passing through water is cut off quickly and though our eyes adjust rapidly to it, the plants of deeper water may suffer a much reduced growth rate. To counteract this, their underwater leaves are usually as large and as thin as possible, presenting the maximum surface to the light. They need little supporting tissue because of the water, and they collapse untidily if you take them out and try to look at them. But float them in a bowl of water again to fully appreciate their structure. Their softness is a disadvantage in the autumn floods and many plants are ripped out of even quite sheltered places. Water plants overwinter often with a stout rhizome or stem sunk in the mud. Some produce heavy, food-storing buds on the principle of the onion. These sink to the bottom and can grow rapidly in spring: no river species depends on its seeds alone.

Animals in Stony Places

The clean hard-water river contains a huge number of animals which come in many different types. When the bottom is of stones and gravel large shrimps will be everywhere. These are orange animals, flattened from side to side and permanently curved into a bow shape. They have rows of appendages on their undersides which they move continuously. They must have good co-ordination for there are two pairs of antennae for feeling, about four pairs of limbs for feeding, five pairs of legs, and various other things at the tail end to help with swimming. They also have a few gills in the middle, because the animal's skin is impregnated with lime which it cannot breathe through. The shrimps are largely scavengers, feeding on dead organic matter and the host of bacteria that grow on it. They often attack other small animals, however, and are fond of soft-bodied mayfly nymphs.

Our shrimp is *Gammarus duebeni* and it probably occurs somewhere on all calcareous rivers and also on many acid ones. In Britain this shrimp is confined to brackish water and *G. pulex* takes over in fresh water, but in Ireland *G. duebeni* is more widely distributed as *G. pulex* was absent here until recently introduced.

Water beetles are also likely to be present in the river along with the mayflies and caddis larvae. There are more sorts of beetle in the world than of any other creature. Many have invaded fresh water, and although few species spend their whole lives there, at least four families are aquatic at most stages of their development. The water beetles are well adapted to the habitat: usually their bodies are streamlined for rapid and easy movement. Their legs, particularly their hind legs, are clothed with hairs which aid swimming. Adult beetles usually depend on atmospheric oxygen. They rise frequently to the surface, and air-holes along their sides trap air, and sometimes their wing cases too. The larvae occasionally are surface breathers also, though often they need less oxygen, being less active, and can make do with the oxygen diffused in the water. The resting pupal stage is spent out of the water, so no problem of respiration occurs.

Some of the beetle larvae you may see have red or yellow bands on their bodies and are 1–2cm long (Haliplidae). This family shows one of the more remarkable features of aquatic organisms the world over, which is that similar animals occur in stony streams in all continents. Visit a stream in South America or China and the chances are you will soon meet a beetle from the Haliplidae, as well as a *Gammarus* shrimp and a water boatman.

Large numbers of snails are common in calcareous rivers. Water snails are smaller than the familiar garden varieties and they have much thicker shells to withstand damage. They form their shell from lime, so they are much more frequent in hard waters than in soft – in water of sufficient acidity the shell would dissolve away, and its functions of protection and ballast would be lost. Snails have developed several shapes: often the axis is more drawn-out than in the land snails, so they have a pointed tip and an obvious spiral, the largest part of which is the living chamber. In contrast to this, the shell of the ram's horn snail is all rolled up in the one plane like a measuring tape.

Some snails have a horny shield with which they can close their shell in adverse conditions. They tend to live in variable habitats in the backwaters and pools, where water levels fall in summer, rather than in the main river channel. Jenkins' spire shell *Potamopyrus* is an interesting member of this group, for though widespread today it was not known in Ireland before 1910. It is a native of Australia and, taken out of its natural community, it has achieved much greater prominence here than it has at home. It occurs in all freshwater habitats from saltmarshes and mudflats to lakes, streams and ponds. Its rapid colonisation illustrates the upstream migration that all animals must undertake to populate our river systems.

Freshwater snails seem to have colonised this habitat from the land like a lot of other organisms. At one time they all breathed air from the atmosphere and some have retained this habit, rising occasionally to the surface to obtain it. They also absorb some oxygen through their soft skins and in well-oxygenated sites they may come up only every few hours. The pond snail *Limnaea* behaves like this and in badly oxygenated water it may be up and down every few minutes. Because of this adaptability it is one animal that survives well in the trying conditions of an aquarium. Most stream organisms demand a high level of oxygen in the water which is difficult to provide artifically, and it is much better to collect creatures from still water.

The snails that live in faster flows do not release their hold to breathe and often they have developed gills within a body cavity filled with water. The freshwater limpet is the extreme case, for this animal remains stuck tightly onto its rock except when feeding. It has lost the spiral shape and looks like a tiny version of it seashore cousin with a little point curved backwards on the shell. Limpets actually grow into a slightly different shape depending on their habitat. In fast flows they are tall and narrow and in slower flows they are flatter and broader.

Life in a Glide

A *glide* is an angler's term for a river flowing smooth and fast over a level bed, a metre or two below. The bed is of gravel, stones and silt, and water crowfoot often forms great trailing clumps. If you follow a stem backwards towards the root it can be 2–3m long. All water plants have an anchorage problem and the crowfoots get over it by being able to root anywhere there is a leaf. Contrast the water-lily, which has a massive root stock sunk in the silt, or the clubrush *Scirpus lacustris* with its branching root stock which forms a type of matting. The underground parts of both plants are actually stems, which can produce new leaves quickly if older ones are stripped away by floods.

The stems of crowfoot are often laden with mayfly nymphs, for example the yellow evening dun *Ephemerella notata* and the large dark olive *Baetis rhodani.* Cased caddis larvae and perhaps free-living ones will be there too. Such larvae often rest with the body bent in a curve. You can see the two limbs at the back end of the body, with which their relations hold onto their cases, and also a large number of hairy gills along its sides. When it swims the caddis bends itself from side to side in a graceful S-shape. The net-spinning caddis *Hydropsyche* may again be here. The nets are spaced out regularly on the sides of stones but are apt to get damaged when the stone is moved. There have been cases of these nets being so abundant as to block the flow in pipes or culverts. *Hydropsyche* is not the only caddis to make a net, although it is perhaps the most skilful builder. *Polycentropus* acts like an underwater spider and makes a snare. It rushes out from hiding when a small animal is swept onto the net.

Although normally associated with river margins frogs are excellen[t] swimmers and can stay submerged for quite appreciable lengths of tim[e]
photograph: Richard Mills

Crayfish occur in many places in a limestone river, but they are rarely seen because they feed largely at night. They are the invertebrate giants of the river reaching 12cm or more in length. They resemble small brownish lobsters, though like lobsters they go bright red when cooked. They have two dangerous-looking claws at the front and a curled-over tail with which they can shoot backwards if disturbed to the safety of a gap between the stones.

Crayfish are carnivorous, eating any small creature that they can catch, such as insect larvae, water snails and sometimes tadpoles. As they grow larger they take more vegetable matter. Pairing takes place in the autumn and the female lays about a hundred eggs. These stick to her body, so she carries them about until they hatch into tiny and vulnerable-looking replicas of herself. The young remain hanging on to the mother for some time and, though fish eat them avidly, their future must be relatively secure.

The glide also contains large numbers of fast-growing trout, lying amongst the plant beds and beside the banks. Their population is high because of the abundance of food and 250 fish can live in a 100m stretch. The growth rate of these trout shows the richness of the habitat: three-year-old fish can be 40cm long and weigh 1kg.

Along the edges of the river a net will reveal many smaller fish.

Having just replenished its air supply this beetle make[s] its way towards the relativ[e] safety of a neighbouring weed bed. Note the great paddle shaped legs coated bristles which help in swimming and give the adu[lt] beetle great manoeuvrability under water.
photograph: Richard Mills

Shoals of minnows, or pinkeens, dart around in unison; sticklebacks hover amongst the weeds, and a stone loach lies on the bottom using feelers on each side of its mouth to search for food particles. Male sticklebacks are highly territorial and they drive all others away from the vicinity of their nests. In breeding colours they have bright red bellies and sides. In a backwater off the main channel may be some gudgeon, small silvery fish that are seldom seen but are quite common.

Otters

Otters eat both fish and crayfish and probably occur in largest numbers on an untouched limestone river. Their maximum density is probably in the order of one family per 5km of river, but you have to be lucky to see this most elusive of Irish animals; even if you do all you are likely to see is a nose above the water, drawing a tell-tale V. To the naturalist studying otters, their redeeming feature is that they leave droppings, called *spraints*, in conspicuous places, using them to mark their territory and to communicate. Whole surveys have been done on the otter based simply on its droppings, and the most recent analysis of its distribution, which showed otters to be present in 95 per cent of all likely rivers, was done without more than two animals being seen. We now know that otters are more frequent in Ireland than in any other country in Europe.

ɪe otter uses its whiskers as vell as its eyes in its search for prey. Its body is ɾeamlined and is insulated by a thick coat of fur. photograph: Richard Mills

Otters are now fully protected, but they used to be hunted, originally for their fur, and later to reduce their presumed competition with fishermen. So it is not surprising that they are shy of people and will sink out of sight rapidly if disturbed. They can stay under water for four minutes and travel at least 400m during this time. The otters you are most likely to see are young ones, rejected by the female at about a year old when she comes to breed again. They are not yet as proficient in hunting as they will be later on and they may be more active by day than the adults.

All otters are opportunists, feeding on the commonest prey wherever they happen to be. They are efficient hunters and can outpace most fish in the water, swimming at a maximum speed of 2–3m a second, which is faster than any salmon. They can do this because, unlike fish, they are warm-blooded. A higher body temperature allows greater activity in cold water. It means that otters usually catch their prey within a few minutes and that most of their hunting trips are successful; so they have time on their hands and, being intelligent animals, they begin to play. Their agility in water must be seen to be believed: they seem totally flexible, swimming upside down or sideways without a check. They use their body and tail to propel them-

'he mink is closely related to the weasel with its curving back and pointed ce and despite this mink's wet appearance, it spends much more time on land than does the otter. photograph: Richard Mills

selves underwater, bending them up and down like a whale rather than a fish. They steer with their paws, but when on the surface they also paddle with them.

Otters have long been blamed for catching the anglers' fish, and sometimes they undoubtedly do this. Yet they do not seek out salmon or trout in preference to other sorts. They are just as likely to live on eels, perch or other coarse fish or on crayfish or frogs. The remains of crayfish are particularly obvious in spraints. The size of the otter's prey may be surprising: in one study the eels eaten were about 30cm long and the fish mostly 10–12cm. The animal itself is a metre or more long, including the tail, so it can take quite large fish, probably when they are confined in a fish pass or a narrow channel.

An otter family is a female with up to three cubs. The males are solitary, but they hold their territory against other males depending on their age and status. The young ones may wander long distances, 10–15km in one night is quite possible and a total range of 60–70km over some months has been noted. The females first breed at eighteen months or two years, but the males are probably older, having to establish a territory for themselves to begin with. The young may be born at any time of the year, which is unusual for our larger mammals. This is because their food supply is almost constant, but in climates with a colder winter than ours most cubs will be born in spring.

The nesting and sleeping chamber is called a *holt,* and it is often set amongst the roots of a tree, perhaps where one has fallen over into the water. If possible the animals enter it under water. Single otters are more flexible and may lie up in reedbeds or in hollow trees. Even families change their holt quite often, as their food supply changes over the season. They are creatures of habit, however, and will often be seen feeding in the same places. They may even enter and leave the water in the same runway.

Mink

One of the most interesting things at the moment about otters is their relationship with mink. The American mink is an introduced animal, which has escaped from fur farms all over Europe and become established in the wild. In appearance it is a bit like a small otter, but it is much less streamlined – its tail for example is distinct from its body. Its hair does not clump together in the spiky way that an otter's does. It will be seen much more on land than the otter: in fact it will be seen much more altogether, being unconcerned and even curious about the activities of fishermen on the river bank. This tameness in a supposedly wild animal is rather disconcerting but presumbly it stems partly from their domestication. It means that they will visit houses and farms in search of food.

Mink and otter do not seem to interact much, which is to be expected as they are different species. However, they both feed on the most available fish, so their food can overlap to the extent of 30–60 per cent. Mink catch smaller fish than otters, and are more catholic in their choice of food, hunting on land and along the riverbanks much more than otters. On some rivers the mink seems to specialise in moorhens, and the relative abundance of this bird is an indication of the numbers of mink. Mink take young mallard and other duck if they get the chance, and also catch field mice, shrews and rabbits, especially young rabbits in spring. It would be interesting to know how often mink eat rats, often the most conspicuous animal on lowland rivers.

In Britain the otter has declined in numbers as the mink has increased, and this has led to the belief that one affects the other. However, in Ireland there is no evidence of a decrease in the otter despite the rapid expansion of the mink since the first escapes in the 1950s. This suggests that there is little real competition between the animals and that the otter's decline elsewhere has been more related to river drainage and the loss of breeding habitat. We are lucky to retain at least some river banks in their natural form and some river waters with abundant fish.

Mayflies

Before leaving the stretch of clean hard water we should examine some of the submerged banks of fine gravel and silt. These are the habitat of the angler's mayfly, *Ephemera danica,* which needs sediment of a certain consistency for burrowing in. *E. danica* is present in large numbers in the Shiven and the Suck in County Galway and the Unshin in County Sligo. The nymph lives in a burrow for security, venturing out at dusk or in darkness to feed. Its burrows are normally a few centimetres deep, but in winter the nymph hibernates in a tunnel that can be 20cm long. Its aquatic stage lasts for one to three years, depending on the supply of its algal food, and in this time it grows up to 34mm in length, the giant in its group. River mayflies are usually larger than those in lakes.

In appearance *Ephemera* is hairier than most other mayflies. Perhaps this is to prevent silt particles from sticking to its body. Its gills also curve up over its back and their beating maintains a water current down the tunnel for breathing. As emergence time approaches the nymphs become more active than usual, and a layer of gas forms between the nymphal and the adult skin. The animal soon becomes buoyant and swims up onto the surface of the water where it stays motionless until the outside skin splits open and the adult, winged form emerges. If it has escaped the attention of hungry fish in this period, the insect then takes off on its new wings and makes for vegetation on the bank. It is a weak flier and its final destination depends mainly on the wind. When it lands it crawls under a leaf, since it has to moult once more before it can mate and lay eggs. It remains on the leaf for a day or more, the immature adult or 'dun' changing into the reproductive adult or 'spinner'.

Spectacular swarms of male mayflies may be seen in afternoon and evening time during hatching. The males flutter up and down in unison above the water's edge and the females fly separately into them when they are ready to mate. After mating the female rests in vegetation for a few hours before making her last flight to lay the eggs. The insect dips up and down on the surface of the water releasing three or four thousand tiny eggs.

It is easy to feel sorry for an animal that flies for only two days and has no jaws for feeding in its adult life. But this is only because we think of the adult insect as the real thing. Such feelings are no part of a mayfly's life. If it could think about it, the nymph would conclude that *it* was the real animal and that the winged phase was a necessary chore to disperse the offspring and make sure that some would survive to perpetuate the species.

This mayfly hatches from mid-May to mid-June and this hatching is the most important annual event for trout fishermen. In many rural areas the appearance of the 'fly' heralds the start of a minor tourist boom as anglers arrive from all directions. The reason for the excitement is simple, as it is only during the emergence period that trout are tempted to feed on the surface and so are susceptible to the artificial fly.

Bird Life in Callows

Lowland rivers in the flattest parts of the central plain seem to lose their sense of direction. They flow little if at all: the Shannon drops only 15m in 215km of its course, the Erne some 3m in 105km. For many miles the rivers resemble linear lakes and the distinction in their wildlife is similarly blurred. Beds of clubrush and reed line their edges and water-lilies float freely on their surface.

Coots, moorhens and grebes cruise amongst the vegetation and herons and snipe stalk their fringes. The pig-like squeals of a water rail rise unexpectedly in the evening and by night you may notice movements of curlews or whimbrels, with their stuttering high-pitched calls.

River valleys are natural lines of communication in the countryside – they have a way of attracting both migrating birds and electricity pylons. Many birds migrate at night and the glint from the water surface must help them to find their way. In the spring the first sand-martins and swallows usually push up the river valleys of the south-east, feeding on the few flies that are already on the wing. The lesser black-backed gull will be seen on rivers and canals making for its inland breeding sites, and occasionally a black tern passes through, pausing to swoop on flies on the reed tops or the water surface.

Just as these first spring birds arrive, our winter visitors are leaving. The waterfowl and waders which are associated with wetlands naturally follow the rivers on their flight to Iceland. White-fronted geese from the Wexford slobs fly up the Slaney valley before branching off to the Barrow and the Westmeath lakes. There is also a noticeable movement of black-tailed godwits up the Shannon system with lapwing, golden plover and a few dunlin. The many thousands of duck that winter on Lough Neagh leave it along the river Bann in many cases. Standing on a bridge along any of these flyways you can appreciate what is happening only by the sound of rushing wings and the occasional call. Almost anything may turn up. Sea-birds quite regularly pass over the chain of lakes and rivers in Connaught and skuas have been seen on the Shannon.

Silt accumulates in the channel of a slow-flowing river in summer, though it is likely to be swept towards the river mouth by winter floods. Banks also fall in and the river carries a heavy load of mud and plant debris. When the rising waters first lap over the banks they lose a certain amount of speed and deposit some of their load. In a natural river, therefore, the level of the banks rises slightly above the surrounding floodplain. In Coole Park, near Gort in County Galway, you can see this clearly, where the river flows through a turlough. Elsewhere it is often masked by past drainage. The floodplain drops fractionally away to the valley's edge. Further marshy ground occurs here, with channels taking water from the valley sides. Whenever a river floods it leaves a fine spread of sediment on the land. Often when the flood has passed you will see the muddied leaves of buttercups standing up through the grass on the callows. This flooding builds up the floodplain in minute layers, secure until a meander chances to cut through them.

Plant Life in Callows

The extreme flatness means that the river edge leads smoothly into marshland, along backwaters and ditches. The soil is waterlogged for long periods so that plant debris does not easily decay and may build up as peat. The courses of the Shannon, the Suck and the Bann are followed by such peatlands for long distances. Plants such as bogbean and marsh cinquefoil show the peaty nature of the ground. The bogbean has huge shamrock-shaped leaves, possibly used by St Patrick and much more suitable for demonstrating the Trinity to a crowd than the traditional clover. Its white flowers are large and feathery, as beautiful as many a garden plant but sparingly produced. Sedges and grasses are everywhere on the callows: sedges have triangular flowering stems, grasses round ones. Sedges may all look alike from a hundred metres but come closer and you will find many differences. Some species grow in large endless beds, some in more discrete clumps and some rise on a pedestal up to a metre above their surroundings. The grey leaves of the bottle sedge are ubiquitous in rich and poor waters, the greener pond sedges also like their feet to be in the water while the almost tree-like tussock sedge often grows among willows.

Along the edges of channels the hollow stems of the water horsetail draw attention to themselves, as do the spongy leaves of the bur reed, often used for resting by dragonflies or the bronze damselflies. The two loosestrifes, purple and yellow, add a bit of colour to the otherwise greenish outlook. If one is on the Shannon or the Bann the three-petalled flowers of the arrowhead or the flowering rush may add their curious beauty. Water mint and gypsywort grow sometimes in and sometimes out of the water, while the amphibious bistort snakes its stems out on the water surface or upwards into the air.

The large bulk of the great water dock, or the heavy heads of bulrushes may erupt out of the reed beds at intervals. Look also at the other end of the scale for tiny duckweeds washed in among the bases of the plants. They are the smallest green plants that float freely in the water, multiplying like algae when conditions are good, but then being swept away by an increase in flow.

Coarse Fish

This is the coarse fish zone, now inhabited by a variety of fish introduced into Ireland. After the glacial period about seven sorts of fish seem to have reached our fresh waters before the sea rose and barred their path. They were char, pollan and shad, all now lake fish, and in the rivers, trout, salmon, eels and sticklebacks. Documentary evidence exists to show that at least seven other species were brought into Ireland in the last four hundred years, and it is probable that another five also were. Species which have done well are pike (*gailliasc* – the foreign fish), perch, bream, tench, roach and dace.

Coarse fish lay sticky eggs which adhere to plants: they are thus adapted to living in the lower and deeper reaches of rivers which trout and salmon can only reach by migration.

They generally feed partly on vegetation and partly on animals. The rudd inhabits weedy river margins, the bream slightly deeper, muddy areas. Both grow slowly and a 30cm specimen could be six to ten years old. Ireland is renowned for the size and abundance of its bream, and they attract many fishermen. In coarse-angling circles the catching of a hundred pounds of bream at one sitting is the dream of every angler and one that comes true fairly commonly in the Erne, Suck and Shannon catchments.

The pike is a large and fast-growing fish which after one year may be 30cm long. Its youthful diet is plankton and invertebrates, but once into its second year the pike eats other fish and grows at an explosive rate. It becomes the top carnivore in the river apart from the otter, able to catch all other fish as well as waterfowl and frogs. It hunts in both shallow and deep water, but may be quite selective in its prey. One population studied used to take trout and perch and to disregard most other fish. Pike may attack large animals such as moorhens, and they have been known to choke on their victims. Male and female fish grow at different rates. All really large pike are females. Males over 4kg are seldom encountered but females can go on to 15–20kg at which weight they are almost a metre long. The giant pike of the Shannon can reach this weight. The pike illustrates a common pattern of growth in fish where increase in length occurs in the fish's early years and increase in weight is astonishing later on. A female pike may be 60cm long when she is four years old but she will weigh only 1kg; at eight to ten years old she may be twice that length but weigh only 15kg.

The perch is perhaps the most handsome of our freshwater fish. It is an olive green colour generally, but has reddish-orange fins. It is a generalised feeder and grows quite slowly. A large perch weighs 2kg and is 25cm long. It is a fish of which there are often large numbers of quite small individuals.

Coarse fish are not widely eaten in Ireland, but both pike and perch are popular elsewhere, and the other species slightly less so. The pike in particular is to be recommended, with its large flakes of meat and its peculiar forked bones.

Invertebrates

Many of the coarse fish take invertebrate food and there are large numbers of some types present. They occur in less variety than in the clearer waters upstream but there are nevertheless plenty to choose from. Cased caddis larvae and snails predominate, several species of mayflies are present too but stoneflies are largely absent.

Stuck onto the stones of riffles or the rivers edge are leeches, segmented animals like worms, which have a sucker at each end of their body. One of the commonest is *Glossiphonia,* which is spotted and slightly translucent. It moves like a looper caterpillar pulling its hind sucker up almost to the front one before releasing this and reaching forwards. Inching along in this way may seem a slow method of progression, but they can escape from a fish pretty quickly. This type feeds mostly on the blood of snails. Their mouths are not strong enough to attempt human skin.

Fish leeches are longer and thinner and they spend a lot of time attached to a stone waving their front sucker in the water. If a fish passes close enough they sink their jaws into it and are carried about until they have had their meal. One of the largest leeches is *Erpobdella,* which lives for several years in the river. It is a fast mover and also has good eyesight: its Latin name is *E. octoculata* or 'eight-eyes'. It feeds on worms and other organisms, which it can swallow whole, and sometimes by night it will leave the river in search of earthworms.

In a rich, slow river a new crustacean will come to notice. It is the water-louse *Asellus,* which looks like a fully aquatic woodlouse. Like its relative the freshwater shrimp, it is a scavenger, feeding on all kinds of decaying organic matter. It can survive lower oxygen levels than the shrimp and may become very abundant if it finds suitable conditions. The mud bottom itself is inhabited by a multitude of worms, some of which live in tubes, and by specialised Chironomid midge larvae – the bloodworms. These look like many other fly larvae – they have few external features, but they are bright red and may be taken for bits of plastic. The redness is caused by haemoglobin in their blood. This substance, which we also have in our blood, has a great affinity for oxygen, and this allows the animal to make use of even the minute quantities found in an organically rich sediment, complete with all its bacteria. The ram's horn snails are similarly endowed.

Peatlands

Rivers in the central plain are often bordered by bogland. They are affected little by intact bogs, because where there is little or no decomposition no nutrients can escape into the drainage water. Water flowing from the bog is noticeably acidic, but this makes little impression on a river loaded with lime. From the centre of a raised bog the ground slopes down to the river, heather and *Sphagnum* moss give way to moor-grass and rushes, and down by the river grassy callows

Overleaf
Here in the Black Valley in County Kerry the endless flow of the waterfall smoothes the roughest rock and gradually wears its wa[y] back up the valley.
photograph: Liam Blake

Fishing cormorants are see[n] on many large rivers but fe[w] remain inland to rest. Here is an adult above one of th[e] Shannon lakes.
photograph: Richard Mills

prevail. White-fronted geese may feed on the callows and fly up onto the bog to roost as they do beside the Little Brosna. If the callows are flooded they may be replaced by Bewick's or whooper swans which graze on submerged grasses and pondweeds.

Uncut bogs are now rare in Ireland, though we have more of them than the rest of Europe put together. Many have been developed for large-scale turf-cutting, usually the production of milled peat. Deep drains are cut through the peat down to the layer of white marl underneath and water soon begins to flow in them. The effect of such drainage is often felt in the river channels, as peat particles escape from the drains to accumulate among the sand and gravel of the river bed. Nowadays the outflow is limited by the use of settling ponds where the heavier particles sink before the water is released, but the remnants of peat banks persist in many rivers. The peat suspended in the water cuts down the amount of light that gets through. Stoneworts *Chara* and shining pondweed *Potamogeton lucens* are the first plants to disappear because of this shading. Species with floating leaves, like water-lilies, can cope somewhat better with it, and the emergent types such as bur reed and clubrush are little affected. Where peat inflow has been severe only the beds of clubrush will survive and they may line the banks below the inflow, even extending because of the silt action. On the opposite bank plants may be largely unaffected.

Peat silt has a stultifying effect on the fauna. It is not a toxic deposit, so few invertebrate species disappear. They just become less frequent because of the lack of algal food and the clogging effect on filter feeders. In severely affected streams trout may cease to breed in the gravel.

Downstream of a developed bog there may be a large inert pile of peat in the middle of the next pool where flow is reduced. The pool may be 3m deep, its edges fringed by tall clubrushes, with the yellow water-lilies further out.

The normal river flora is beginning to reassert itself and between the submerged leaves of the lily the tolerant Canadian pondweed is caught. You may have seen beds of this neat water plant in upriver sections of the hard water river growing close and regular on the bottom. Before 1836 you could not have seen it anywhere in Europe, for this was the year it first appeared at Waringstown in County Down, and was later recorded in a pond in Dublin. It spread rapidly from then on because, like many water plants, fragments of its stems caught in the river silt can grow into new plants. By 1860 it had spread to France and Belgium and since that time it has gone further, reaching today into Sweden and southern Siberia and also in the southern

Previous page
Here in the Gap of Dunloe, County Kerry the black and grey patches on these rocks are lichens that are wetted for different periods each year.
photograph: Liam Blake

An underground waterfall in the Burren region of County Clare taking the rainfall down through the porous rock.
photograph: Liam Blake

hemisphere to Australia and New Zealand. The plant showed the typical symptoms of a new introduction at the start, usually reaching a peak of abundance in any new habitat in five to seven years. In the beginning only female plants were recorded, so all the spread must have been by vegetative means. Male plants, or at least male flowers, remain unknown in Ireland even now, but have been seen in Britain. Gradually the Canadian pondweed seems to have lost its early vigour and become assimilated into our flora. It no longer clogs small water courses and presumably is subject to some natural, but as yet unknown, control.

How this plant gets around without seeds and even how the seeds of others disperse is a perennial mystery. Plants can obviously colonise new areas downstream from where they are but how do they move upstream or to new catchments? Wandering birds and mammals may explain some cases of transport: waterfowl have certainly been seen with plant fragments sticking to their feet and feathers and some seeds may be eaten but still retain viability. Single otters travel such great distances that they must be one of the most important agents of dispersal. Nowadays fishermen and boat traffic are effective methods and barges on our canals and rivers are implicated in the distribution of several plants. The minute waterwort *Elatine hydropiper* spread from Lough Neagh down the Lagan canal to Belfast and also along the Newry canal where it may still be found. Arrowhead, *Sagittaria*, now grows all along the Shannon, but also in the Grand and Royal canals and on most of the Barrow. The flowering rush *Butomus* has not reached the Barrow but occurs in a few places on the Liffey.

Weirs and Millraces

Weirs and artificial rapids rejuvenate the lowland river. The turbulence they create allows aquatic mosses to reappear and oxygenates the water for the more demanding types of animals. Below the weir water crowfoot may again appear in the channel and in shallow backwaters close to the bank a skin of diatoms may grow rapidly across the mud in spring or summer.

Millraces and the edges of rich rivers are the places to look out for kingfishers. They need bankside bushes or drift wood to perch on, and they dive straight down into the water to seize small sticklebacks or minnows. The water must be clear enough to see the fish, and during spates when the river is charged with mud they may seek out ditches and mill ponds in which to hunt. Kingfishers are not uncommon, but until you have learned their high-pitched fluty call, often uttered in three notes, you will not see them often. It is surprising how inconspicuous such a brightly coloured bird can be if the weather is dull and the kingfisher is some distance away. Their fast, straight flight is another thing to look for as they cross the river into a willow bush.

Kingfishers nest in a tunnel that they dig in a vertical face of sand or soil. Usually it is on the river bank itself perhaps on the eroding face of a bend or where a tributary joins the main stream. They can occasionally use the soil lifted by a fallen tree or a ditch bank that becomes dry in summer. Their nesting hole develops quite a fishy smell after a time, and the droppings of the young may be obvious below it. It is also single, unlike the tunnels of the sand-martin, which at least in the north and west can nest in the same situations. Kingfishers are much less tolerant of hard winters than are dippers, and many move downstream during the winter appearing in estuaries and other places where a few reeds or a wire fence offers them a perch.

Some of these lowland waters are the adopted home of the roach, a small rudd-like fish that before 1968 was confined to the Cork Blackwater and the Fairywater in Tyrone. Both of these are well-documented introductions, but since that time the fish has established itself in many other catchments, thanks to the activities of anglers using line bait or to intentional stocking. The roach has invaded the Erne, Shannon, Corrib and Liffey systems and their possible impact on native trout in the midlands and west is giving cause for concern. At the moment they are caught below weirs on parts of these rivers. They don't do active damage by eating other fish, but they come to dominate a catchment in a very short time. Their egg-laying capacity is exceptional. At one spawning site on the Annalee it was estimated that over 270 million eggs were laid in an area of less than 100 square metres. Every introduced species seems to go through a phase of rapid increase before it settles down and fits into the natural community. But the roach's behaviour makes it unlikely ever to be accepted in the trout angler's river, because it will take bait intended to catch trout.

Estuaries

A hard-water or soft-water river loses much of its identity when it flows into the sea and meets an immense volume of nutrient rich water. For a time the fresh water floats on the salt water, but the turbulence caused by winds and currents gradually mixes the waters together. The estuary is the dumping ground for river sediments: mud from the river and sand and shells from the seashore are arranged and rearranged in banks and mudflats. Each rising tide lifts fine sediment and transports it upstream: when the river reasserts itself on the ebb it is brought seawards again.

This muddiness is an obvious feature of a tidal river and occurs higher up the river than the salt influence. The regular rise and fall of the water also keeps the banks soft and slippery helping them to grow immense sappy plants of water dropwort or angelica. The smell of algae, possibly the tubular *Enteromorpha,* hangs in the air and the willow trees grow large. The grey and the white willows, and the crack willow with its brittle twigs, are common, withies or osiers are planted everywhere and on south-eastern rivers the almond willow also appears. These sites have seldom seen drainage machinery and their bankside vegetation flourishes like nowhere else. Above their banks there may be artificial dykes built to prevent spring tides from inundating agricultural land. Estuary muds are rich in the extreme and form good soils, so good that it is worth reclaiming them from the sea.

Reeds often fringe the estuary; on the Slaney and the Suir they form immense beds that swish in the wind. The bottom mud is rich in invertebrate life which provides food for several sorts of fish. Salmon and sea trout smolts spend some time in the estuary adjusting to salt water before their sojourn in the sea. The flounder comes into the estuary and even into fresh water to feed and the twaite shad breeds in some of the south coast rivers before returning to the sea to grow. Bass may also move in and out on the tide, and there is also the constant movement of cormorants and the drifting of the gulls.

5 Rivers and us - then and now

In almost every part of Ireland there are stretches of river which seem unspoilt and apparently quite free from the influence of human beings. You have probably walked along a river bank which is slightly elevated so that the view is alternately the wide stretch of callow land away from the river and the long reach of the river itself, possibly punctuated by a weir or other feature. The masonry bridge in the distance is so hallowed by time that it seems to have been there for ever. But as they travel down from the mountains to the inhabited lowlands, rivers come more and more under the influence of human activity, and have done for centuries.

The elevated bank which makes walking so easy may well be the spoil heap thrown up during one of the many drainage schemes hand-excavated by starving labourers in the famine years. The weir, whose sole function now appears to be to improve the landscape, may be all that remains of a long-forgotten attempt to capture fish or to provide motive power for a local mill. The bridge, known for centuries past as the old bridge, was once the brash new bridge that replaced a succession of wooden bridges, which had in turn replaced a ford or a ferry. Ditches, banks, side branches, weirs and bridges, the presence of stands of butterbur, *Petasites hybridus,* or balsam, *Impatiens,* the absence of trees or the occurrence of planted poplars and willows all tell of the ceaseless interaction of people and rivers over the centuries.

Human activities generally have an enriching or *eutrophying* effect on rivers and few waters have retained the nutritional level they had when people first reached our shores. First of all, people cleared the forest; this gradually changed flow conditions in the river. Trees intercept a good proportion of the rain that falls on them, they catch it on their leaves and it evaporates back into the air without ever reaching the ground. They also build up a thick layer of dead leaves and branches on the soil and a good layer of roots below. Both of these absorb rainfall, which is used by the trees themselves and also by soil life. A beech forest evaporates 60 per cent of the year's rainfall through its leaves alone, certainly in a continental climate.

So what form did our rivers take in post-glacial times? Flows must have been much lower through most of the year and run-off after rain much slower than today. Is river flooding quite a modern phenomenon brought about by forest clearance? This is definitely so in many parts of the world today, but in Ireland it is more likely that winter rains falling on a saturated forest floor have always run off quite rapidly into rivers, causing bankfull discharges, erosion and flooding. But the frequency of flooding must have been increased by human activities and spates in the April–October period must now be an occasional factor in river life that was not there before.

We cannot know for sure how human activity so long ago has affected our rivers, but we know something about the relationship between people and rivers in more recent times.

The History of Fishing Weirs

Fishing weirs were originally formed by driving stakes into the bed of the river and then intertwining wattles between them to form a barrier of basket-work. A gap was left, either in the centre of the stream or at one of the banks, with some form of staging from which fish were caught by netting the gap. In the estuaries of tidal rivers such fishing weirs were usually built pointing downstream and the salmon, which tend to move up and down stream with the tidal flow, were thus trapped on the ebbing tide. Similar fishing weirs were constructed in bays around the coast.

Contemporary chroniclers refer to the defeat of the Danes by Brian Boru at Clontarf in 1014 as the battle of the salmon weir. It is probable that many more local disputes, some leading to blows and possibly to bloodshed, arose from time to time in connection with the operation of such simple weirs on our inland rivers and tidal waters. Fishing rights then as now were a sure way to warm the blood.

Simple stake-and-wattle weirs were still used as late as the nineteenth century. Richard Hoare in his *Tour of Ireland,* published in 1807, refers to such a weir on the Munster Blackwater:

> Ballinatray. The weirs on the Blackwater are not (like those at Limerick and many other places) flood weirs extending across the whole river but are fished only during the latter half of the ebb. The wings are staked and wattled where there is least current, so as not to impede the navigation, and are only as large as half the flood. . . Where they meet in an angle the fisherman has a seat upon four posts where he holds the net and on feeling the salmon strike collects his and draws it into the boat.

This description of the construction and operation of a stake-and-wattle weir could be applied to the fishing weirs constructed in great numbers over the centuries in both tidal and non-tidal rivers. Fish traps were also constructed to catch eels who migrate in the opposite direction to salmon and sea trout.

In recent centuries stake-and-wattle salmon walls were largely replaced by loose stones and finally by masonry walls. In these later weirs the gap was constructed in the form of a box or crib which the salmon could enter through a gap in a pair of converging walls at the downstream end. They remained trapped because of a grating on the upstream side. A few fishing weirs of this type are still being operated in Ireland. One of the most famous of these was the Lax weir in Limerick, which has a continuous fishing history of over seven hundred years. By the nineteenth century the Lax weir consisted of fifty stone piers and twelve boxes or cribs. When the Shannon was harnessed for electricity in the present century the Lax weir became ineffective and it was replaced by the present Thomond weir of reinforced concrete. In recent years the fish pass at the Thomond weir has been supplemented by a pass at the site of the power house.

From earliest times there were provisions that an open gap should be left either permanently or temporarily to ensure an upstream movement of salmon to the spawning grounds adequate to maintain the stock. The battle between those seeking to ensure the future stock of fish and those interested only in personal short-term advantage continues to the present day.

The ownership of fishing rights and of milling rights at various times is a good indicator of social change in Ireland, as ownership changed from communal property to partnership among neighbours to feudal rights to modern ownership by limited company. Up to the time of their dissolution, most of the Irish monasteries operated substantial fisheries and in most cases leased out the fishing rights which were more extensive than required even for such a centre of fish consumption as a monastery.

Harnessing the River – From Watermills to Electricity

Legend tells us that the watermill was introduced into Ireland in the second half of the third century by King Cormac Mac Airt in pity for his beautiful bond maiden whose customary task it was to grind corn with a hand quern. There is ample evidence of Irish watermills, many of them connected with monasteries, as early as the sixth century. By the seventh and eighth centuries, corn mills operated by water power were in common use throughout Ireland. Over the following thousand years most of the Irish mills were operated by small water wheels set on their sides. These rotated in a horizontal plane as a result of the impact of the running stream on a series of radial blades. These small mills served an individual *clachán*, or extended family settlement, and were an integral part of the subsistence agriculture of their time. A number of them were still in operation in the nineteenth century but few traces of them remain today. These smaller horizontal mills were distinguished in common Irish speech from larger vertical mills by the description *muileann tón le talamh* (mill with backside to the ground).

Some of the Irish mills must have been quite substantial from an early date. A number of the annals relate the story of the death at the hands of the Leinstermen in AD 651 of the sons of King Blathmac. Wounded, Denagh and Conall took refuge inside the mill housing, only to be crushed when their pursuers forced the woman in charge of the sluice to set the mill in operation. It is clear from the enumeration of the eight parts of a mill in the account of the Brehon Laws in the Senchus Mor that these larger mills were fed by a headrace and a mill pond, the pond providing storage for a period of continuous operation.

The importance of water power for a medieval monastery and the degree of mechanisation based on it can be appreciated from the following quotation from a life of St Bernard of Clairvaux.

> The river enters the abbey as much as the wall acting as a check allows. It gushes first into the corn-mill where it is very actively employed in grinding the grain under the weight of the wheels and in shaking the fine sieve which separates the flour from the bran. Thence it flows into the next building, and fills the boiler in which it is heated to prepare beer for the monks' drinking, should the vine's fruitfulness not reward the vintners labour. But the river is not yet finished its work, for it is now drawn into the fulling-machines following the corn-mill. In the mill it has prepared the brothers' food and its duty is now to serve in making their clothing. This the river does not withhold, nor does it refuse any task of it. Thus it raises and lowers alternately the heavy hammers and mallets, or to be more exact, the wooden feet of the fulling-machines. When by swirling at great speed it has made all these wheels revolve swiftly it issues foaming and ruffing as if it had ground itself. Now the river enters the tannery where it devotes much care and labour to preparing the necessary material for the monks' footwear; then it divides into many small branches and, in its busy course, passes through the various departments, seeking everywhere for those who require its services for any purpose whatever, whether for cooking, rotating, crushing, watering, washing or grinding, always offering its help and never refusing. At last to earn full thanks and to leave nothing undone it carries away the refuse and leaves all clean.

We tend to forget that a key factor in the mechanisation of technology in European history was the monastic system, and that the main motive for this mechanisation was to provide more time for prayer and for intellectual work. The great monasteries were located on rivers, not because of the fishing, but because of the need for water power and water transport.

The horizontal water wheels were gradually replaced by larger wheels rotating vertically. These vertical water wheels originally turned under the action of the stream beneath them but later the construction of a millrace allowed the water to enter the wheel at any point in its circumference. In the seventeenth and eighteenth centuries the mills increased in size and there was a great controversy about the relative efficiency of undershot wheels, in which the power was produced by the impact of the flowing water near the bottom of the wheel, and high breast overshot wheel in which the power was due to the weight of the water falling in a closed bucket. It would appear that the type most common in Ireland were breast wheels in which water was fed above the level of the main axle into buckets. The controversy between the advocates of undershot water wheels and overshot water wheels was only settled in favour of the latter when the eminent civil engineer John Smeaton carried out tests on small scale models of various designs.

In the eighteenth century most Irish water wheels were built of wood and, though they were slow moving, some of the parts wore out quickly. In the nineteenth century the all-wooden water wheels were replaced by cast iron water wheels with better transmission gearing and better control devices. Some of these mills are still to be seen in various parts of the country and a few are still in operation.

In the nineteenth century the competition between mill owners in urban areas for the use of the potential power of any available head of water gave rise to much dispute and litigation. In some cases the execution of a combined scheme for drainage, navigation and water power by the Board of Works enabled a rational plan to be adopted. Thus in Galway the scheme under which the Eglington canal was built between 1848 and 1858 to allow navigation between Lough Corrib and the sea also made provision for the improved operation of over thirty mills. These depended for power on the fall of 8–9m on the Galway river between Lough Corrib and the sea. At the time of this scheme, the watermills in Galway included fifteen flour mills, five corn mills, three breweries, two distilleries, three malt mills, two coach mills, two foundries, a marble mill and a paper mill.

With the discovery of electricity some of the water wheels were adapted to power local lighting schemes but the advent of a national electricity grid led to the abandonment of most of the older water wheels in the past half century. However, the last decade has seen the reinstatement of local water power on the basis of modern turbines of a size suitable for the small flow and small head available.

The installation of major hydroelectric works has transformed the rivers on which they are built, both by changing the flow and by creating artificial lakes. The effect on the flow of the Shannon was not so marked because this river always had the large natural storage of Lough Allen, Lough Ree and Lough Derg, and the same is true of the Erne. The Erne hydroelectric scheme was designed in consultation with the authorities in Northern Ireland, so the resulting scheme has both produced power for the Republic and improved the drainage conditions for farmers in Northern Ireland.

In the case of the Liffey and of the Lee, the creation of artificial reservoirs not only changed the regime of the flow of the river but replaced a river ecology by a lake ecology. In the case of the Lee scheme, the reservoir behind Carrigadrohid dam flooded out part of the Geragh, which was a unique area of interconnected channels and small vegetated islands. In County Donegal a diversion weir was built on the River Clady and the flow diverted by a headrace canal nearly two miles long to the Gweedore river, so that power could be developed by a combination of the flow of the Clady and the fall on the Gweedore.

Crossing the River

The improvement of natural crossing points on rivers and their replacement by bridges reflects a fascinating story of technological progress, which is reflected in many Irish placenames. We find a succession of terms such as *áth* (a ford) and *clachán* (stepping stones), *ces* (a causeway of hurdles) and *droichead* (a bridge).

The Brehon Laws prescribed that the chief builder (*ollamh saoir*) should be efficient in the erection of bridges. Bridges were apparently commonplace in Ireland by the ninth century. Until the seventeenth century most of the bridges were wooden beam structures. Some thirty years ago, the remains of a wooden bridge dating from about AD 1600 were excavated during arterial drainage works on the river Cashen in Kerry. From the reconstruction of this bridge and from manuscript sketches of other bridges, it is clear that these bridges would be easily destroyed in a major flood or by determined military action, and the annals often record such destruction.

It is often stated that stone arch bridges were not built in Ireland until the twelfth century. It is possible, however, that the monks built small stone bridges whose design would not call for any greater expertise than that shown in building small churches. The earliest surviving stone arch bridges are the Monks Bridge on the de Vesci estate near Abbeyleix and the Castle Street Bridge in Trim, County Meath. Both would appear to date from the thirteenth century and have a number of small arches of varying shape with massive intermediate piers. These early stone bridges offered a serious obstruction to the flow of the river in time of flood. Borman in his book on *The Bridge* has written of the relation between the bridge components and the river as follows:

> These two, the pier and the arch constitute our bridge. The history of their quarrel is the history of the development of bridge form. In the pier the bridge conquers the stream; in the arch it yields, an act of homage.

The growing homage of the bridge to the river in the form of increased waterway can be seen in the bridge across the river Barrow at Leighlinbridge dating from 1320, which, though later widened, still retains its original profile. There is evidence of an increased hydraulic efficiency resulting in an enhanced appearance in the eighteenth-century bridges spanning the Barrow at Graiguenamanagh and spanning her sister river the Nore at Kilkenny. The contest between bridge and river still persists.

River Transport

From time immemorial boats and rafts have been used on our rivers. Many dug-out canoes have been found in the course of river drainage and bridge construction. Such boats continued to be made until the seventeenth century, when the forest clearances made the large trunks necessary for dug-outs difficult to find. Boats were then made out of planks, but they resembled dug-outs, without a definite stem or bow. Derivatives like these are still in use on our south-eastern rivers and on the Shannon, the Erne and the Bann, where the eel fisheries gave them a new lease of life. The river cot is tapered, but cut off square at each end. Some were propelled by poles, especially at river crossings, but oars were also common, set on a wooden pin or between two pegs.

In the eighteenth and nineteenth centuries there were numerous efforts to develop further the system of inland navigation throughout Ireland. In 1700 the cost of transporting goods by water was less than half of transportation by road. A barge could carry fifty tons on a canal or thirty tons on a river, compared with a wagon, which could carry only two tons on a well-made road. In 1715 the Irish parliament passed an Act

> to encourage the draining and improving of the bogs and unprofitable low grounds and for easing and dispatching the inland carriage and conveyance of goods done under this Act

And the new act in 1729 established the 'Commissioners of Inland Navigation for Ireland'. Half of the cost of works was to be met by tolls and local contributions and half by a tax on such luxury items as carriages and sedan chairs.

Under the 1729 Act the Newry canal, which was the first summit-level canal in these islands, was constructed to join the Newry river to the upper Bann and hence to Lough Neagh. The economic justification for the canal was to facilitate the shipping of the coal being mined on the shores of Lough Neagh to Dublin where there was a relatively large market for coal. After a number of engineering and legal difficulties, the canal was finally opened for traffic in March 1742.

In 1755 work was commenced on the Grand canal and on the Barrow, Boyne and Shannon rivers, which were to be linked by a comprehensive scheme of inland waterways. In designing a canal provision has to be made for supplying it with water. The Grand canal required a 10km feeder from Pollardstown Fen in Kildare for its summit level. The Grand canal was almost entirely excavated as a new cut, but much of the work on the Barrow, the Boyne and the Shannon consisted of the improvement for navigation purposes of the existing river channels – and the by-passing of rapids by short canals and locks. The linking of Lough Neagh with Belfast in 1790 through the Lagan canal, and with the Erne in 1840 through the Ulster canal, and the subsequent linking of the Erne with the Shannon in 1860 through the Ballinamore and Ballyconnell canal completed the network of interconnection between the main Irish rivers. Though many of the elements of this network were commercial failures from the beginning – and none is any longer of commercial interest – the network is being increasingly availed of today as a leisure amenity.

Draining the Land

Most of the rivers of Ireland have been affected, at least over part of their length, by works of arterial drainage undertaken to provide adequate outfalls for land drainage or the development of bogs. The Shannon Commission was established in 1831 to consider both the navigation of the Shannon and its tributaries and also the practicability of draining the land subject to prolonged winter flooding and less severe but more damaging summer flooding. In 1842 the experience gained on the Shannon was used as the basis of a general Act for the improvement of drainage, navigation and water power on rivers generally. Between 1842 and 1846 surveys and designs were completed and works commenced under this Act in six catchment areas. The work was enormously expanded during the famine years. Between May and October 1846, surveys were completed and plans and estimates prepared in 101 districts. Between 1846 and 1851 works were carried out in over a hundred districts. Later works were carried out under the codes established by successive enactments in 1863, 1925 and 1945.

Arterial drainage is the most severe form of management for the river. It is carried out to speed the flow of water to the sea, preventing floods and lowering the water level in the channel so that field drains can empty into it. The means to achieve these ends are to widen or deepen the channel and to eliminate obstructions to water flow. Speeding up the river flow is liable to increase erosion on bends, and therefore the channel may be straightened or canalised and the banks protected with stonework or concrete. Bringing machinery in to do this work and to dispose of the soil necessitates the removal of trees, so the appearance of the river as well as its habitats for river life are drastically altered.

Once drained, flow in the river changes in character. The extremes of high and low flows are often intensified. There is less storage capacity in the catchment, so run-off is more rapid and the river reaches its peak discharge more rapidly after a rainstorm. Likewise low flows become lower as drained agricultural land cannot supply a steady flow. In an extreme summer a drained river may not even be able to water cattle. Most aquatic organisms are adapted to quite a narrow range of flow conditions and if these are exceeded one way or the other in a drained river they cannot survive. Also water depth is normally reduced by drainage and there is less physical space available to water plants and animals.

Every river channel is subject to periodic scouring by flood waters and to the gradual transport of rock and soil from the mountains to the sea. Thus they may recover from this most drastic form of scouring. Where the draglines have not gone below the gravel layers in the channel, recovery seems to be very rapid. Beds of water crowfoot again expand and larger numbers of invertebrates occur. Exceptional numbers of trout may also be found but when examined it turns out that they are all young fish of one to three years. There are probably more fish present than before drainage and over a thousand may inhabit a kilometre stretch. But the fish are slow-growing and in poor condition because the dredging has done away with the deeper pools in which larger fish used to find shelter. Abundant food is available for the younger trout in the form of mayfly nymphs and fly larvae and a population explosion takes place in the fish. However, few survive long after their first spawning, and the large trout that made the Maigue or the Robe great trout rivers for the angler have vanished for ever.

Trees are also removed for drainage so there are no shady patches, no undercut banks held together by old roots and none of the nooks and crannies needed by otters or kingfishers. The uniform conditions allow a smaller range of plants to grow and the clubrush, helped by the direct sunlight, is often the most conspicuous. Indeed it can flourish to such an extent that it becomes a nuisance. It is often the main reason for further maintenance on the channel with all its attendant disruption. Where the gravel layer has been removed a bed of rich silt may be exposed. This is unsuitable for trout and becomes colonised by minnows, sticklebacks and gudgeons. It provides an ideal habitat for resting pike and these may be found here waiting to make a sortie into neighbouring trout zones.

The drainage of rivers has been going on for a long time, but it is mainly since the advent of heavy machinery that its effects have been so serious. When the work was done by hand and when the river provided water power for mills and other works, only short sections of the channel were cleared out and recovery was much more rapid. Along many lowland rivers today you will see the old millraces and dilapidated weirs that used to divert some of the river flow. Often the channels themselves are almost dry or only have pools of standing water in them, but some still carry a smooth flow and a few still operate watermills.

Taking Water from the River

Rivers and lakes have been the main source of usable water in Ireland up to the present and will remain an important factor even when groundwater resources are drawn upon in order to meet further industrial development. Whereas in earlier centuries the abstraction of water from springs and from rivers was only a small fraction of the available flow, this is no longer true. Ireland has available about four times more water per person than the more densely populated countries of Europe, but this is not the case in the eastern region of this country where one-third of the population is concentrated.

The amount of water required to maintain life has been estimated to be 1–3 litres a day, but the social minimum required for the prevention of disease is about 100 litres per person per day. To this must be added the ever-increasing amount required to maintain agricultural and industrial production. In the east of Ireland over one million people depend on an average run-off of 7000 litres per person per day, which is certainly adequate. The run-off in this eastern region during drought periods, however, can fall to below 400 litres per person per day, which would limit normal domestic and industrial use. So it seems that before very long we shall have to contemplate large schemes of water transport from parts of the country where there is plenty of water.

About four-fifths of all water abstraction in Ireland at present comes from rivers or lakes. The remaining one-fifth, derived from groundwater, is mostly used by individual farms and group supply schemes or by creameries and food processing plants.

Modern Farming

Agriculture is the most pervasive of human activities, and it is carried out in some form on all the good land in a river catchment. Its direct impact on a river remained small until recent times. Forage was saved as hay rather than as silage, animals were housed on straw rather than on slatted floors, and crops were grown on the natural fertility of ploughed-up grassland rather than with chemical fertilisers. Intensification has changed all this in the last forty years, and it has also created an unending demand for the deepening or draining of rivers.

Along much of the course of a river you will see silage being made twice or three times in early summer. Silage is fresh grass pickled in acids that are produced naturally by bacteria in the absence of air. During its production, water equivalent to about half of its original weight is lost, and this trickles away from the clump in the first few weeks. In some farms this effluence is collected in slurry tanks and later spread on fields; in others it seeps into ditches and drains causing luxuriant growth of nettles or other plants. In many cases it reaches a water course where it has immediate detrimental effects.

Wastes are decomposed as part of the natural cycle which begins with green plants, the basic producers of organic matter. Everything else – all animals as well as the fungi and bacteria which are not green – depends on green plants for their food, either directly or indirectly. Each of the 'consuming' organisms takes its food in a slightly different way. Grazing animals, like mayfly nymphs and aquatic snails, scrape off the skin of algae from rocks and plants; many fly larvae wait for the plants to die and then chew their way through the general detritus on the river bed, dead tree leaves and wood, even the droppings of other animals; fish and water boatmen are carnivorous, catching small plant-eating animals; in their turn they are themselves eaten by otters and herons – the top carnivores on many rivers.

No organic material in nature is wasted, all form the food of one organism or another. If it is not eaten by an animal it is used by those great scavengers, the fungi and bacteria, which complete the breakdown of everything into simple substances like carbon dioxide, nitrate or phosphate. These are then again available to plants for growth.

It follows from the presence of such an army of consumers that any organic waste that gets into the river is immediately seized upon by some organism and used for its food. The consumer will grow and multiply, and it may swamp out organisms that were formerly present or make conditions unsuitable for them. Thus pollution is often marked by multiplication of one or a few forms of life which reach incredible numbers.

Although it is a liquid, silage effluent is loaded with organic matter, largely the acids produced by pickling. It is soon discovered by bacteria already present in the water in small numbers. They multiply enormously and form a stringy hair-like growth which, before people knew of its composition, was called sewage fungus. Other bacteria and fungi are mixed in with the main organism, *Sphaerotilus*, and they all require oxygen from the water. Sometimes all the available oxygen is removed from the water course resulting in the death of many stream invertebrates and also of fish if they cannot escape it. The sewage fungus persists as long as there is material on which it can grow, and it may line one bank of a stream for a considerable distance. But gradually the water becomes oxygenated again in riffles or on weirs. Algae and other plants release oxygen into the water and things return to normal.

Nevertheless, silage effluent is probably the most serious pollutant generated by farming, but because it is produced for only a few weeks it is easy to shrug off its importance and to take no precautions over its disposal. It is a highly concentrated effluent; in fact it has an oxygen demand two hundred times that of ordinary domestic sewage. The amount of silage made today is about 15 million tonnes, and it is still growing by half a million tonnes a year. The effluent from this pile of silage amounts to about 8 million cubic metres, equivalent to the yearly flow of a small river. Not all this reaches our waterways, but it can be seen to be a threat to the health of many rivers.

Silage pollution is temporary – it occurs over a month or two in the summer – though it is when river flows are normally at their lowest. Animal wastes are produced all year round depending on how many animals are housed. They are collected in a watery slurry, which is stored in a tank until it can be spread on the land. With well-drained land such spreading is possible on and off through most of the year. Where the soil type is heavy it retains the slurry for long periods, and any more sprayed on the surface will run off into ditches and streams.

These soils are characteristic of the north midlands which are our main pig-farming area. Pig slurry is an exceptionally rich substance, containing large amounts of phosphate, nitrate and potassium. It is a valuable fertiliser which promotes plant growth. The trouble is that too much growth happens too quickly in aquatic habitats and results

Overleaf
The river Boyne seen here flowing peacefully near Trim in County Meath has witnessed prehistoric burials and St Patrick's paschal fire, Celtic monasteries and Norman castles, Williamites and Jacobites battling for the throne of England and mo[re] recently prosperous agriculture and highly mechanised bog development.
photograph: Liam Blake

Two anglers fly fishing for salmon near the weir on th[e] Corrib at Galway. The streamlined salmon is exceptionally well adapted to dealing with strong currents and when running can often be found holding station in the most turbulen[t] white water pool.
photograph: Liam Blake

again in the deoxygenation of the river water.

The waste from any animal establishment is rich in organic matter and in ammonia or other nitrogenous substances, potassium, phosphate and sulphate – all common constituents of living things. Algae respond immediately to such nutrient chemicals, and a plant such as blanket weed, *Cladophora,* may live up to its name and grow rich and dense across the channel. *Cladophora* has a stringy feel to it, unlike the sliminess of many other green algae. As the sewage fungi and bacteria get to work on the more solid matter, they release more nutrients and give the algae another surge of life. Animals will come to seek shelter in this moving weed. If they eat algae or bacteria they will find abundant food and, during the daytime, abundant oxygen. At night, however, the growth of algae and the consequent release of oxygen no longer occurs, though the animals and plants continue to use it. Just before dawn there may be so little oxygen in the water that many normal stream animals suffocate, being unable to rise to the surface and breathe like their land-based relatives. Fish kills may occur then, and it is also the time to see the dedicated biologist taking measurements of the oxygen content of the water.

Such pollution does not 'kill off' a stream however. Looking at the habitat from the point of view of a bacterium, a tube worm or a water-louse, feeding conditions become excellent and the population can explode. From other points of view, though, things are not so good. Most fish are killed or, more likely, they desert the affected stretch of water, the variety of plant and animal life is much reduced, the water becomes cloudy and it may have a noticeable smell, and the sewage fungus and other slimes make paddling in it unsavoury.

Slurry pollution may occur anywhere, but it is especially pronounced in the drumlin areas of the Erne and the Inny catchments. The effect on lakes and groundwater is well known, but the over-enrichment may affect the rivers just as severely. The multiplication of algae in the lakes (algal blooms) is the most noticeable thing, and the water may appear as if green paint has been poured onto it. Before the transport scheme which allowed the export of slurry away from Lough Sheelin the Inny brought some of this load successively into Lough Kinale and Lough Derravaragh which responded in the same way. Sampling the Inny showed it to carry a hugely increased load of plankton, mostly blue-green algae.

The impact of so many blue-green algae from polluted lakes on river life wherever it occurs is not perhaps what would be expected. As many sorts are distasteful or poisonous, there may be no great multiplication of animals feeding on them.

Previous page
This eighteenth-century
nasonry bridge at Clara in
County Wicklow has
withstood many a flood on
e flashy Avonmore River.
photograph: Liam Blake

rplus water flows over the
p of a canal gate in Dublin,
where the V-shaped design
uses the water itself to
maintain a good seal with
the masonary of the canal
ock – a device introduced
by Leonardo da Vinci.
photograph: Liam Blake

The Town on the River

All our large rivers flow through towns of one sort or another and as they approach the coast they meet centres of population and industry more frequently. Pollution by sewage is a feature of all of them in places. The effects of food-processing industries, like creameries and sugar factories, are perhaps more serious, though they are much more localised.

The guiding principle in the disposal of sewage is to remove it as fast and as hygienically as possible from where people live. This concern goes back to the days when sewage was a health hazard and epidemics of typhoid and other diseases were rife. Farmers are now coming to see slurry as a potential resource, which should be recycled into the soil to keep up its fertility, but our culture has never viewed sewage in this light.

The fertilising or enriching effect of sewage is noticeable below an outfall, and the release of small quantities is seldom harmful to a river. Self-purification takes place as the organic matter is broken down and its nutrients absorbed to the benefit of plants and animals. However, 'small quantities' is the important point. In many cases sewage is directed into streams that are quite unable to cope with the load, especially in summer.

The most common sewage treatment works is a simple settlement tank which catches the solids and decomposes them to some extent and releases the liquid from the top. If the tank is too small the sewage does not have sufficient time in it to separate properly and a greyish cloudy outflow results. This flows as a discrete band in the river and it can often be followed back to its source.

The effluent will usually contain the swaying strings of the sewage fungus, which will appear at intervals downstream growing on stones or branches caught in the river. As we have seen, the organisms that normally break down organic matter need oxygen to do it, and their growth and multiplication reduces its level in the water. If they use up all the dissolved oxygen they put themselves out of business.

But decomposition does not stop there. There are also *anaerobic* organisms, which flourish only in the absence of oxygen. The substances these produce during growth are quite different, and it happens that we find three of them particularly smelly – most probably because we are oxygen-demanders ourselves. Nitrogenous matter, such as proteins, give rise to ammonia and hydrogen sulphide (rotten egg gas) while carbon-containing compounds are broken down into methane. Waxes and oils are produced naturally, which float on the water surface. Hydrogen sulphide is the most offensive gas, and it creates (with iron) the blackish colour in the mud of most polluted streams. Bubbles of it accumulate on the bottom and every so often they rise to the surface in clods of mud. Ammonia is much more toxic to fish and other organisms and more so in hard waters than in soft. In a clean natural river there may be only 0.01 parts of ammonia per million of water because it immediately reacts with oxygen to form nitrates. In polluted waters levels of 0.5 ppm ammonia cause trouble at least to fish and there have been records of more than 200 ppm in some effluents.

Where the sewage enters a stream the bottom mud is usually anaerobic during summer and left to those bacteria that can take it. At other times of the year the most tolerant invertebrates invade it from slightly downstream. These are the tube worms, *Tubiflex*, and the bed of the stream will often be covered in their red, swaying filaments. If you touch the bottom with a stick the redness disappears suddenly as the animals withdraw into their tubes. A few of these worms exist naturally in the richer muddy rivers, but they are especially favoured by low levels of oxygen and abundant food. They may occur in enormous numbers, up to 400,000 in one square metre. The mud is also likely to contain huge numbers of bloodworms, the larvae of the midge *Chironomus*, and dense clouds of the adult insects may swarm over such waters on summer evenings.

Moving downstream away from the pollution source the fauna becomes richer in species, though not in total numbers. One of the ways of quickly assessing pollution is to look at the relationship between these two things. The more polluted site will have high numbers of one or two organisms, the less polluted smaller numbers of a selection of species. The effluent becomes mixed with other river water and is diluted and its constituents begin to be broken down. Large colonies of the water-louse *Asellus* are characteristic, as are concentrations of snails, *Limnaea*, and the ram's horns with their dependent leeches.

There are certain to be blackfly with their bulbous hind ends. Blackflies, *Simulium*, are not part of our everyday experience here, unlike in North America, but you may see the adults in spring, small-bodied ordinary looking flies often with white legs. Their eggs are laid in jelly-like masses on stones and plants at the edge of streams, and the larvae wriggle into the stream and attach themselves to rocks or water plants. They spin a little pad of sticky silk onto the surface and then hold onto it with a ring of hooks at the tail end. When a larva wants to move it spins another pad on the surface and grasps it with its jaws. It then moves its tail end up and repeats the process so it resembles a leech in its movement. It also has a safety line like a spider and can pull itself back upstream if it is temporarily dislodged.

The blackfly larva feeds by sieving particles of food, especially bacteria, from the water with two comb-like bristles on the head. It stretches up outside the boundary layer and smoothly flowing water. Because turbulence would wash the food out of its sieves the larva chooses a site with quite a fast but regular flow. When it is fully grown it spins a case onto a rock that resembles a tiny cone on its side. In this it pupates and when the adult is ready to emerge, gas accumulates between the skin of the adult and that of the pupa. The pupal skin splits and the adult fly floats to the surface enveloped in a silver bubble of air. The wings are ready to function straight away, and it takes flight immediately it reaches the surface.

Blackflies can be found everywhere in both rich and poor streams, but on days in May they can bring on a frantic rise of trout. Huge swarms appear on the Boyne and its tributaries in May. They also attract bats to feed, and there are some bats that habitually hunt over water. Ireland has seven species of bats and you may often see the Pipistrelle as it roosts in old mill buildings and houses. The Daubentons bat seems to be much rarer. It roosts under bridges.

Along with the writhing masses of blackfly larvae and the leeches there may be other things attached to the small stones. These are flatworms which can be passed over at first as blobs of jelly. If you wait for them to recover from the shock of being lifted out of the water, they will elongate to a centimetre or so and glide slowly over the surface, their movement a cross between a snail and a leech. Unlike a leech they have no segments or divisions on the body and they are in fact very primitive animals, well-suited to a life of eating small animals and detritus particles that come their way.

This is the sort of place to find a few really large brown trout if there are pools for them to lie in. They may be quite young fish that have thrived on the limitless diet of invertebrates. But, feeding on the bottom and living in such conditions their flesh is soft and white, and their taste far from pleasant. There may be some record-breakers among them, however, 2–3kg fish being quite common.

The large populations of invertebrates below a sewage outfall are sustained by the organic particles in the sewage and the bacteria that settle on them. Their effect is to liberate simple chemicals from the sewage that can then be taken in by plants and used for their growth. So below an outfall you get a peak in abundance of sewage fungus and other bacteria, then a peak in algae and a more gradual rise of higher plants. Higher plants grow along the edges of the water and the reed canary grass and clubrush may grow to a spectacular size, waving up to 2m above your head. The sedge warbler, like most birds, is not fussy about aesthetics. If the habitat is right and the food plentiful it will nest in any reed bed and you will hear its staccato churring song by dusk and day. Willow bushes nearby may provide song-posts for the reed bunting and grasshopper warbler.

The first submerged plant to appear below a polluted zone is usually the fennel pondweed *Potamogeton pectinatus*, one of a group of grass-leaved pondweeds with the finest of leaves and occasional heads of a few flowers or seeds. Its flowering spikes just break the surface, for the pollen is dispersed by floating on the water. Having flowered, they sink back in, so they are seldom seen unless you haul the plant in to shore. Below the pondweed zone the water crowfoot returns with vigour and then water starwort and the more normal river flora.

Mute swans feed on river plants, especially the narrow-leaved pondweeds, and they are well known on every large river. One of the things they need is a sufficient length of water to take off from. Flying is obviously necessary for the bird's safety but it is also used to mark the territory of a nesting pair, the singing of the wingtips replacing a vocal song. If there are people around swans soon become accustomed to them, and will quickly come to take bread, but they are far from gentle when nesting and it is unpleasant to be caught at close quarters with an angry adult. In open situations the nest is often conspicuous, a huge pile of vegetation with the adult sitting securely on it. It can be hidden, however, in reeds or on a river island, and the first that is known about it is the adult female leading a brood of five or more downy cygnets with the male taking up the rear.

Organic pollution cures itself in time and if no further inflows occur a river may be almost back to normal five or ten kilometres below a town. Unless it is a soft-water river – in which case the nutrient conditions will be totally altered – its life comes again to resemble a natural river. That stretch below the town, however, causes a blight on the river. It may be unsightly or offensive to smell, it cannot be used for swimming or for supplying fish to eat and it may also block the spawning migrations of salmonid fish to higher parts of the river.

Several inland towns have built sewage treatment works in recent years and new ones are also under construction. The idea of such a plant is to condense all the natural purification that would occur in the river channel into various tanks and ponds within the works. If the residence time of the effluent is long enough the bacterial and some of the invertebrate stages are gone through on land before the effluent is released. Most of our treatment works do not have a 'plant stage' to absorb the nutrients out of their effluent, so algae may become incredibly abundant. The night-time deoxygenation they cause may block the movement of fish and kill all but the most resistant animals, such as bloodworms and water-lice. Some treatment plants reoxygenate their effluent to try and get around this problem.

Surveys in recent years have indicated that out of 6900km of river channel, 1000km show some loss of water quality. In the case of about 130km of channel, this loss of quality is serious. Most of the badly affected stretches are on streams or small rivers.

The potential for waste treatment must be seen as one of the values of our rivers. In the natural course of events they deal with a substantial input of organic matter and, provided they are not overloaded, they will usually retain both health and beauty. They will then be useful for other things, filling their rightful place in the appearance and economy of the countryside. Compared to the desolate canalised rivers of much of lowland Europe, which flow treeless, straight and murky to the sea, our rivers remain as treasures. In most places they are still valuable for water supply and agriculture, for fishing, boating and other recreations.

Index

INDEX OF PLANTS AND ANIMALS

English names of plants and animals are used in this book for the most part. Occasionally Latin names are given, usually to distinguish plants and animals with the same English name. Latin names are only indexed when they appear in the text, so if you find only one instance of a plant or animal under its Latin name in the index you may find more if you look it up under its English name.

INDEX OF RIVERS, LAKES, CANALS